CRTD-Vol. 43

An ASME Research Report

# HISTORY OF LINE PIPE MANUFACTURING IN NORTH AMERICA

PREPARED BY
J. F. KIEFNER
KIEFNER AND ASSOCIATES, INC.

E. B. CLARK
COLUMBIA GAS TRANSMISSION CORPORATION

FOR THE
GAS PIPELINE SAFETY RESEARCH COMMITTEE

UNDER THE DIRECTION OF
STEERING SUB-COMMITTEE, HISTORY OF LINE PIPE MANUFACTURING
IN NORTH AMERICA

REVIEWED BY
INDEPENDENT PEER REVIEW COMMITTEE

THE AMERICAN SOCIETY OF MECHANICAL ENGINEERS
UNITED ENGINEERING CENTER / 345 EAST 47TH STREET / NEW YORK, NY 10017

**DISCLAIMER**

This report was prepared as an account of work sponsored through the American Society of Mechanical Engineers (the Society) Center for Research and Technology Development by Columbia Gas Transmission Corporation, El Paso Natural Gas Company, Gas Research Institute, Research Committee on Gas Pipelines Safety, and Washington Gas and Electric Company (collectively referred to herein as the Sponsors).

Neither the Society, nor the Sponsors, nor Kiefner & Associates, Inc. (referred to herein as the Sponsoree), nor any financial contributors or others involved in the preparation or review of this report, nor any of their respective employees, members, or persons acting on their behalf, makes any warranty, expressed or implied, or assumes any legal liability or responsibility for the accuracy, completeness, or usefulness of any information, apparatus, product, or process disclosed, or represents that its use would not infringe upon privately owned rights.

Reference herein to any specific commercial product, process, or service by trade name, trademark, manufacturer, or otherwise does not necessarily constitute or imply its endorsement, recommendation, or favoring by the Society, the Sponsors, the Sponsoree, or any financial contributor or others involved in the preparation or review of this report, or agency thereof. The views and opinions of the authors, contributors, and reviewers of the report expressed herein do not necessarily reflect those of the Society, the Sponsors, the Sponsoree, or any financial contributors or others involved in the preparation or review of this report, or any agency thereof.

ISBN No. 0-7918-1233-2

Library of Congress Number 96-79838

## Steering Committee History of Line Pipe Manufacturing in North America'

Edward B. Clark, Columbia Gas Transmission Corp.
Jon O. Loker, P.E., Consultant
Kenneth C. Peters, Southern Natural Gas Company

## Peer Reviewers, History of Line Pipe Manufacturing in North America

Ted Bruno, Metallurgical Consultants, Inc.
Ernest Jonas, Consultant
Angus MacDonald, Consultant
E. L. Von Rosenberg, Materials and Welding Technology, Inc.

## Gas Pipeline Safety Research Committee

F. Roy Fleet, Chair, Consultant
Wesley B. McGehee, P. E., Vice Chair, Consultant
Jeffrey C. Langlinais, Secretary, Phillips Petroleum Company
Robert F. Allen, ARK Engineering and Technical Services
Robert C. Becken, Pacific Gas and Electric
Bennett M. Brooks, P.E., Brooks Acoustics Corporation
Steven C. Coleman, Kansas Corporation Commission
Cesar De Leon, P.E., U.S. Department of Transportation
James C. De Voe, P.E., Consultant
Robert L. Dean, Consultant
A. J. Del Buono, A. J. Del Buono Inc.
Carl P. Hendrickson, Northern Illinois Gas Company
Jerry F. Holwerda, Consumers Power Company
Sam Kannappan, GDS Engineers, Inc.
Melvin F. Kanninen, MFK Consulting Services
Jon O. Loker, P.E., Consultant
Jerry L. Lucas, Southern California Gas Company
John C. Murphy, Johns Hopkins University
William F. Quinn, P.E., El Paso Natural Gas Company
Michael Zandaroski, Minnegasco

# LIST OF PUBLICATIONS
## ASME CENTER FOR RESEARCH AND TECHNOLOGY DEVELOPMENT

| Volume No. | Title |
| --- | --- |
| CRTD-1 | *Consensus on Current Practices for Lay Up of Industrial and Utility Boilers*, 1985, 28 pp. Book Number H00336 |
| CRTD-2 | *Research Needs in Thermal Systems*, 1986, 221 pp. Book Number I00212 |
| CRTD-3 | *Goals and Priorities for Research in Engineering Design*, 1986, 164 pp. Book Number I00230 |
| CRTD-4 | *Pressure Systems Energy Release Protection (Gas Pressurized Systems)*, 1986, 425 pp. NASA Contractor Report 178090 |
| CRTD-5 | *Thermodynamic Data for Biomass Materials and Waste Components*, 1987, 496 pp. Book Number H00409 |
| CRTD-6 | *Analysis of the Feasibility of Replacing Asbestos in Automobile and Truck Brakes*, 1988, 136 pp. Book Number I00264 |
| CRTD-7 | *Hazardous Waste Incineration: A Resource Document*, 1988, 192 pp. Book Number I00266 |
| CRTD-8 | *Research Needs in Dynamic Systems and Control – Volume 1: Strategic Initiatives and Opportunities*, 1988, 58 pp. Book Number I00275 |
| CRTD-9 | *Research Needs in Dynamic Systems and Control – Volume 2: Acoustics and Noise Control*, 1988, 48 pp. Book Number I00276 |
| CRTD-10 | *Research Needs in Dynamic Systems and Control – Volume 3: Control of Mechanical Systems*, 1988, 40 pp. Book Number I00277 |
| CRTD-11 | *Research Needs in Dynamic Systems and Control – Volume 4: Machine Dynamics*, 1988, 37 pp. Book Number I00278 |
| CRTD-12 | *Research Needs in Dynamic Systems and Control – Volume 5: Nonlinear Dynamics*, 1988, 32 pp. Book Number I00279 |
| CRTD-13 | *Peer Review Thermal Energy Storage Program*, 1989, 46 pp. Available from CRTD, Washington, D.C. |
| CRTD-14 | *The ASME Handbook on Water Technology for Thermal Power Systems*, 1989, 1900 pp. Book Number I00284 |
| CRTD-15 | *Research Needs and Technological Opportunities in Mechanical Tolerancing*, 1990, 51 pp. Book Number I00298 |
| CRTD-15-1 | *Selected Case Studies in the Use of Tolerance and Deviation Information During Design of Representative Industrial Products*, 1992, 156 pp. Book Number I00329 |
| CRTD-15-2 | *Mathematization of Dimensioning and Tolerancing*, 1992, 38 pp Book Number I00337 |
| CRTD-16 | *Engineering Fluid Mechanics Workshop Report*, 1990, 114 pp. Book Number I00299 |
| CRTD-17 | *ASME Research Committee on Corrosion and Deposits From Combustion Gases Seminar of Fireside Fouling Problems*, 1990, 183 pp. |
| CRTD-18 | *ASME Ash Fusion Research Project*, 1990, 11 pp. Book Number I00307 |
| CRTD-19 | *Achievements in Tribology*, 1990, 173 pp. Book Number H00269 |
| CRTD-20-1 | *Risk-Based Inspection – Development of Guidelines: Volume 1, General Document*, 1991, 155 pp. Book Number I00311 |
| CRTD-20-2 | *Risk-Based Inspection – Development of Guidelines: Volume 2, Part 1, Light Water Reactor (LWR) Nuclear Power Plant Components*, 1992, 156 pp. Book Number I00321 |
| CRTD-20-3 | *Risk-Based Inspection — Development of Guidelines: Volume 3, Fossil Fuel-Fired Electric Power Generating Stations Applications*, 1994, 178 pp. |

# LIST OF PUBLICATIONS
## ASME CENTER FOR RESEARCH AND TECHNOLOGY DEVELOPMENT

| Volume No. | Title |
| --- | --- |
| CRTD-21 | *Hazardous Waste Destruction by High Temperature Incineration,* 1991, 21 pp.<br>Book Number I0266A |
| CRTD-22 | *Mining Workshop for Nuclear Waste Cleanup,* 1991, 145 pp.<br>Book Number I00314 |
| CRTD-23 | *The Use of Decision-Analytic Reliability Methods in ASME Codes and Standards Work,* 1993, 120 pp.<br>Book Number I00351 |
| CRTD-24 | *ASME/Bureau of Mines Investigative Program on Vitrification of Residue From Municipal Waste Combustion Systems,* 1994, 132 pp.<br>Book Number I00373 |
| CRTD-25 | *Installation of Plastic Gas Pipeline in Steel Conduits Across Bridges,* 1993, 64 pp.<br>Book Number I00353 |
| CRTD-26 | *Lubrication Technology for Advanced Engines – An Assessment of Industrial Needs,* 1993, 104 pp.<br>Book Number I00352 |
| CRTD-27 | *1993 International Forum on Dimensional Tolerancing and Metrology,* 1993, 324 pp.<br>Book Number I00360 |
| CRTD-28 | *Research Needs and Opportunities in Friction,* 1993, 48 pp.<br>Book Number I00365 |
| CRTD-29 | *Research Guidelines for Aluminum Product Applications in Transportation and Industry,* 1994, 240 pp.<br>Book Number I00366 |
| CRTD-30 | *Tribology in Manufacturing Processes,* 1994<br>Book Number I00374 |
| CRTD-31 | *An Evaluation of the Cost of Incinerating Wastes Containing PVC,* 1994<br>Book Number I00377 |
| CRTD-32 | *Hazardous Waste Incineration: What Engineering Experts Say,* 1994, 17pp.<br>Book Number I0266B |
| CRTD-33 | *Final Report From the Aluminum Industry Workshop,* 1994, 273 pp. |
| CRTD-34 | *Consensus on Operating Practices for the Control of Feedwater and Boiler Water Chemistry in Modern Industrial Boilers,* 1994, 38 pp.<br>Book Number I00367 |
| CRTD-35 | *A Practical Guide to Avoiding Steam Purity Problems in the Industrial Plant,* 1995, 40 pp.<br>Book Number I00383 |
| CRTD-36 | *The Relationship Between Chlorine in Waste Streams and Dioxin Emissions From Waste Combustor Stacks,* 1995, 716 pp.<br>Book Number I00385 |
| CRTD-37 | *Screw-Thread Gaging Systems for Determining Conformance to Thread Standards,* 1996, 72 pp.<br>Book Number I00391 |
| CRTD-38 | *Assessment of Factors Affecting Boiler Tube Lifetime in Waste-Fired Steam Generators: New Opportunities for Research and Technology Development,* 1996, 116 pp.<br>Book Number I00393 |
| CRTD-40-2 | *Risk-Based Inservice Testing: Development of Guidelines — Volume 2,* 1996,142 pp.<br>Book Number: I00388 |
| CRTD-Vol. 43 | *History of Line Pipe Manufacturing in North America,* 1996, 292 pp.<br>Book Number I00396 |

# TABLE OF CONTENTS

## TABLE OF CONTENTS (CONTINUED)

# TABLE OF CONTENTS (CONTINUED)

# TABLE OF CONTENTS(CONTINUED)

## List of Tables

## List of Figures

## TABLE OF CONTENTS (CONTINUED)

SECTION A


INTRODUCTION

This document was prepared for and with the support of the Gas Pipeline Safety Research Committee of the American Society of Mechanical Engineers. Its purpose is to provide pipeline operators with historical data on line pipe, so that they will be able to operate their pipelines, particularly the older ones, with greater confidence in their safety and reliability.

The document is comprised of four major sections. The first explains the manufacturing processes that have been and are being used to make line pipe. The second presents tables by type of pipe listing the manufacturers of line pipe, past and present, in North America. At the end of this section some techniques for identifying unknown pipe samples are presented. In the third section the API line pipe specifications as they have evolved since 1928 are reviewed. The fourth section is a glossary of terms frequently associated with line pipe manufacturing.

Besides being extremely grateful to the Gas Pipeline Safety Research Committee, the authors would like to acknowledge the support of several individuals and, in some cases, their companies. First, and foremost we are thankful for the support of the following companies and individuals who have contributed figures, photographs, and historical documents without which we could not have completed this work.

- Mr. Don Kemper and Moody Totrup International, Inc. for giving us access to the old API specifications dating back to 1928.

- Mr. Scott Metzger and Stupp Corporation for providing numerous drawings and photographs depicting ERW manufacturing processes.

- Mr. Roy Fleet, Natural Gas Pipeline Company, Mr. Greg Morris, Panhandle Eastern Corporation and Mr. Bud Ludwick, East Ohio Gas Company for providing old purchase agreements and mill reports on line pipe orders.

- Mr. W.B. Smith, retired from Youngstown Sheet and Tube Company, for access to his voluminous files on line pipe manufacturers.

- USX Corporation for permission to use drawings and photographs from various U.S. Steel and National Tube publications.

- Wheeling-Pittsburgh Steel Corporation for permission to use a figure from a Pittsburgh Steel Catalog.

- Napa Pipe Corporation for permission to use figures from its brochure.

- Berg Steel Pipe Corporation for permission to use two figures from its brochure.

- Iron and Steel Engineer for permission to use figures from Mr. G. W. Shuetz's 1976 article "Current Trends in Seamless Tube Mill Design" and several figures from two editions of the "Making, Shaping and Treating of Steel".

- Taylor Forge Corporation for permission to use a picture from one of its bulletins.

- Fried, Krupp AG (Hoesch-Krupp) for permission to use figures from Hoesch publications.

Besides these specific inputs, it is to be acknowledged that virtually every manufacturer who was contacted provided us with at least a current brochure. This includes most of the manufacturers listed in the tables of manufacturers.

We are particularly indebted to the following individuals who reviewed the draft of the document and provided many helpful suggestions.

Mr. E. L. Von Rosenberg, President, Materials and Welding Technology, Inc.

Mr. T. V. Bruno, President, Metallurgical Consultants, Inc.

Mr. E. A. Jonas, Consulting Metallurgical Engineer

Mr. Angus MacDonald, Consultant

Last, but not least, the authors are grateful to Ms. Leota Alwine for her dedication to finding obscure references, deciphering old texts, compiling enormous amounts of data and carrying out voluminous correspondence related to our efforts.

# SECTION B

# PROCESSES FOR MANUFACTURING LINE PIPE

# 1.0 HISTORICAL OVERVIEW

Prior to 1812 pipe and tubing for various uses was hand-made from wrought-iron plate by heating, bending, lapping and hammering the edges together. In 1812 an Englishman named Osborne invented machines to do much the same thing in a process known as "hammer lap-welding". Later, in 1824 and 1825 came the process known as butt welding or furnace butt welding. Still later followed the development of continuous lap welding in the 1840's. The first extruded wrought-iron seamless tubing appeared in the 1836 with improvements arising in 1840 and 1845. It remained a costly process, however, and seamless tubing was little-used until the invention of rotary piercing in 1886. The invention of the Bessemer process for making steel in 1856 made possible the first butt-welded and lap-welded steel pipe in 1887. By 1900 most pipe was made from steel by either butt welding (1/8 to 4-inch diameter) or lap welding (2 to 12-inch diameter).

The development of line pipe manufacturing methods as we know them today began in the 20th century. Experiments with electric resistance welding (ERW) began in the period between 1910 and 1920. By 1924 the essence of the low-frequency ERW process was embodied in the "Johnson" process. Flash-welding and direct-current ERW are important variations of the basic low frequency technique. High frequency ERW pipe was introduced in 1955 and by 1970, it had largely superseded low-frequency welding. By the late 1920's the processes for making large-diameter seamless pipe had been developed. Electric fusion (submerged arc) welding of line pipe was attempted in the 1930's, reappeared in the 1940's and became the standard method for making large-diameter (24-inches and larger) pipe by the 1950's.

Readers and users of this document will undoubtedly be most interested in those methods of manufacturing which account for the bulk of the pipe which has been installed in pipelines in the 20th century. These processes are:

Furnace butt-welded pipe (including continuous-weld pipe)
Furnace lap-welded pipe (including hammer-weld pipe)
Seamless pipe
Electric resistance welded (ERW) pipe
Flash-welded pipe
Submerged-arc-welded pipe
Spiral-weld pipe.

Descriptions of these processes are presented on the following pages.

## 2.0  FURNACE BUTT-WELDED AND LAP-WELDED PIPE

Butt-welding of wrought-iron pipe was invented in Great Britain by James Russell in 1824.[2-1,2-2]  In this method a plate was heated and bent to form a cylinder by butting the white-hot edges together.  The tube was formed by round grooves in a hammer/anvil arrangement followed by reheating the crude tube and passing it through a round groove in a rolling mill and over a mandrel supported in the opening between the rolls.[2-2]  This method was quickly overshadowed by the invention of "bell" welding in 1825 by Cornelius Whitehouse (also of Great Britain).[2-2]

### The Processes

### Butt-Welded Pipe by the "Bell" Method

The following description of this now-obsolete process comes mostly from two sources: "The Making, Shaping and Treating of Steel[2-2] and National Tube Company's 1935 catalog[2-3].  Pipes in sizes from 1/8 to 3 inches were made by this method usually in 22-foot lengths.  The process started with "skelp" a hot-rolled strip of steel the length, width, and thickness of which were tailored to a specified finished pipe size.  The edges of the skelp were slightly beveled as shown in Figure 2-1(a) so that they would meet squarely when brought together.  The leading edge of the skelp was trimmed to a "V" shape and turned up slightly to be gripped by "tongs" that pulled it through the "bell".

The bell where pipe forming and welding was effected is shown in Figure 2-1(b).  Prior to welding the skelp was charged into a reheating furnace until it reached the proper temperature for welding (1450°C, 2642°F).  The entire piece was heated uniformly to this temperature in a special furnace, hence it is often called "furnace" butt-welded pipe. The V-shaped end was gripped by tongs, and the tongs and skelp were drawn through the bell as it came out of the furnace. The bell had limited movement so that as pulling started the bell was actually in the furnace. It was pushed into the furnace with the tongs immediately before the gripping of the skelp. Then, as the whole assembly, skelp, tongs, and bell were drawn out of the furnace by a draw chain, the bell hit a stop, the skelp passed through it thereby being forced into a tube with the edges coming together to bond. As the last part of the skelp passed from the bell, the bell then dropped and was picked up for reinsertion of the tongs to start the next piece.

In addition to the basic process described above, associated steps, or in some cases, alternate processes were necessary to create finished pipe. For the larger sizes, above 1-inch for example, the skelp was passed through two successive bells rather than one. Apparently, the first bell created a partially-curved non closed tube while the full tube was "welded" by the second bell. Pipes of all sizes were "sized" after welding, that is, still hot, they were reduced in size by being forced between two rolls grooved to form perfect round pipe of nearly the size desired (see Figure 2-1(d)). The pipe was then allowed to cool to a temperature somewhere in the range between 800°C (1472°F) and 1000°C (1832°F) at which time it was passed through a "scale-breaker". The scale breaker consisted of a series of three pairs of rolls (typically two sets horizontally-oriented and one set vertically-oriented) shaped like the sizing rolls shown in Figure 2-1(d). These rolls changed the size of the pipe only slightly; the main purpose was to break up the mill scale which had formed on the outside diameter and inside diameter surfaces and thereby remove it. This operation was supposed to be performed at a low enough temperature such that the scale was brittle while the pipe was still hot enough to deform readily.

After scale-breaking, the pipe was allowed to cool further. It was then passed through curved and skewed straightening rolls (see Figure 2-1(e)). Other operations referred to as cold finishing may or may not have been performed. In all cases the ends were cropped by cold sawing and each pipe was subjected to visual inspection. Pipe could be rejected for crookedness, bad seams, laminations, blisters and rough spots if they could be detected visually. Cold straightening was permitted as was cutting out defective segments to leave short but acceptable pieces. Threading was done if required. Finally, each piece was hydrostatically tested at an internal pressure ranging from 700 to 2200 psig. At a special hydrostatic test bench movable heads pressed and held to the pipe between water-tight packers while the pipe was pressurized for a few seconds. During the hydrostatic test, if required, the pipe was "jarred" by an air hammer to indicate its soundness. Upon completion of the test the test water was removed by elevating the pipe suddenly so that the water would carry off any loose scale. The pipe was then weighed and bundled with others for shipment.

## Continuous Butt-Welded Pipe

Butt-welding as described above was a relatively slow process because the skelp for each piece had to be loaded into the furnace, heated, gripped and pulled through the bell. The furnace was charged with many pieces of course, but production rates seldom topped 100 feet per minute. In 1923 the Fretz-Moon[2-4 to 2-8] process for making continuous butt-welded pipe was introduced. The process starts with previously-hot-rolled, coiled skelp much like an ERW mill. The coil is unrolled, leveled and looped through a preheating furnace. It then enters a long (150-foot) furnace which heats the entire strip of skelp to welding temperature 1340 to 1430°C (2450 to 2600°F). A residence time of only 30 seconds is needed so welding speeds in excess of 300-feet per minute are possible. The skelp is fed through to start a run but thereafter it is gripped and pulled through the furnace. Upon emerging from the furnace, the edges of the skelp are descaled by jets of air. This blast of air through heat of oxidation also raises the temperature of the edges above the welding temperature 1350°C (2460°F). The skelp passes thereafter through 6 pairs of grooved rolls (arranged in three sets each set having a vertical-axis pair and a horizontal-axis pair). The first set forms only a partial tube, the second set brings the edges into intimate contact. Welding takes place here. The third set supplies most of the traction for pulling the skelp through the mill. Continuity of the process is maintained via a traveling loop of skelp which allows a long enough delay to permit the end of one coil to be flash-welded to another.

Continuously butt-welded pipe is then sized in much the same manner as furnace butt-welded pipe (see Figure 2-1(d)). Unlike furnace butt-welded pipe it is cold-straightened. It is also subjected to hydrostatic testing before leaving the mill. By 1966 continuous butt-welding had supplanted entirely the old furnace butt-welding process. It is still used today to make pipe for a variety of purposes in sizes ranging from 1/2 to 4-inches. It is still recognized by the API Specification 5L as an acceptable process for making line pipe although it is limited in API to one grade (Grade A25). Other specifications such as ASTM A53 permits a "Type F continuous butt-welded material with a 30,000 psi minimum yield strength, and discussions are underway to have a 35,000 psi minimum yield strength grade.

## The Lap-Weld Process

Lap-welded pipe including the hammer-weld process occupied an important segment of pipe-making history.[2-1] Until the advent of large-diameter seamless and ERW pipe in the late 1920's lap-welding was the only technique used to produce line pipe in sizes over 3 inches. The first great wave of pipeline building in the 1920's utilized primarily lap-welded pipe. This includes the major pipelines built to carry natural gas from the Great Plains areas to mid-western urban markets. So more than one of the large gas transmission companies which survive today started in business with pipelines comprised of lap-welded or hammer welded pipe.

Lap-welded pipe manufacturing is described in detail in References 2-2 and 2-3. The sizes of lap-welded pipe ranged from 1¼ to 30 inches. Since the primary interest is in the larger sizes the discussion below shall apply primarily to sizes 8 inches and up. But the main difference between the small and large diameter processes was associated with how the skelp was made rather than how skelp was made into pipe.

The feedstock for a lap-welded pipe mill was in the form of plates with sheared or scarfed edges trimmed to length so that each plate would make one joint of pipe. The lengths for single joints were 22 to 26 feet. Double joints were 44-feet long. The width of each plate was approximately the final circumference of the pipe. In the first step of the pipe-making process, the plate was charged into a bending furnace where it was heated to temperature in the range of 700 to 800°C (1300 to 1470°F). If the plate edges had not already been scarfed at the rolling mill they were scarfed at this point. Scarfing refers to the preparation by rolling of tapered edges which constituted a bonding surface much larger than the wall thickness (as shown in

Figures 2-2(a) and 2-2(b)).  The heated plate or skelp for pipe up to 16-inches outside diameter was drawn by tongs through the "bending" die which deformed it into a cylinder.  A mandrel was suspended over the die to start and to control the bend.  For pipes of 16-inches outside diameter and larger the bending was done hot by a three-roll pyramid.  The top roll was cradled in the other two so as to force the plate into a cylindrical shape around it.  Once the "can" was formed it had to be withdrawn end-wise off the top roll.  ("Can" is a term applied today to the cylindrical shells formed for submerged-arc welding (SAW) to make SAW pipe.  Whether or not the same term was ever applied to the cylinders prior to lap welding is not known).

After being bent to the form of a tube each piece was reheated to a temperature above 1350°C (2460°F) because mill scale remains molten above that temperature and can be forced out of the bondline region during welding.  If the mill scale were allowed to solidify the welding would have been of poor quality or nonexistent.  Since carbon steel with a carbon content of 0.10 percent begins to melt at about 1455°C (2650°F) the temperature range for welding was narrow and it was monitored visually.

When a can was ready the welder positioned the stationary welding ball shown in Figure 2-2(d) between the welding rolls (see Figure 2-2(c)).  A heated can from the furnace was then pushed, lap-up, over the nose of the ball and into the rolls.  The entire cylinder was then pressed between the rolls and the ball forcing the lapped edges down to the thickness of the pipe and bonding them together.  The ball, which was attached to a long "welding bar" was then withdrawn and removed from the bar.  It was quenched in water and reinstalled on the bar in preparation for welding the next can.  This welding cycle required about one minute per 22-foot piece.  For boiler tubes a second welding pass was made after a brief reheating following the first welding pass.

If the weld appeared sound the next step was sizing followed by straightening. The sizing rolls were similar to the welding rolls shown in Figure 2-2(c) and the straightening rolls were similar in principle to those used for butt-welded pipe (see (Figure 2-1(e)) but were much longer and more skewed.

After the welding, sizing, and straightening operations had been completed, the pipes were slowly rolled on a cooling bed. The larger sizes were sprayed with water to speed up the cooling. At the end of the cooling bed, the pipes were visually inspected. The ends were cropped by cold sawing and the pipes were again inspected for defects. Finally, each piece was subjected to hydrostatic testing as described in the butt-welded pipe section. Standard pipe was tested at pressures between 600 and 1000 psig but test pressures as high as 3000 psig were used for some products.

**Hammer-Welded Pipe**

Lap welding was limited to a pipe size of about 30-inches. At that size it became difficult to maintain the shape of the bent can in the lap-welding furnace. Therefore, for pipe diameters ranging from 20 to 96 inches, the process of hammer-welding was developed. Pipes in the range of 20 to 30 inches were made by either lap welding or hammer welding.[2-3] Individual joints of hammer-welded pipe could be made in lengths up to 30 feet. Pipes requiring skelp wider than the mills could roll were made with two hammer-welded seams.

The skelp for hammer-welded pipe was formed into cans in a set of pyramid rolls. Usually this was done cold but for very thick materials the skelp was heated prior to forming. The forming was carried to the point where the edges overlapped. The degree of overlap was sufficient to assure a bond zone relatively much wider than the thickness of the plate.

The weld was made in successive segments by heating the over lapping edges both from the outside and inside surfaces with gas flame jets as shown in Figure 2-3. When the proper temperature was reached, a supporting anvil was placed under the inside diameter surface and the outside diameter surface was struck repeatedly with a hammer until the edges of the skelp had been forged flush with the original wall thickness as shown in Figure 2-3. After being welded the pipe was annealed to remove stresses and sized in a set of rolls much the same as those used to size lap-welded pipe.

## Quality Factors

### Reliability of the Seams

It was recognized that the seams in butt-welded, lap-welded, and hammer-welded pipe materials tend to be less reliable than the parent pipe material. The internal pressure design formulas for both liquid pipelines (ASME B31.4) and gas pipelines (ASME B31.8) specify "joint" factors of 0.6 and 0.8, respectively, for butt-welded (including continuously-welded) and lap-welded pipe materials. Hammer-welded pipe is not mentioned in either of these codes nor was it ever included as a material in the API Specification 5L, but it is assumed to be covered as a type of lap-welded pipe. In contrast, the joint factor is 1.0 for seamless, modern ERW pipe, and DSAW pipe, in recognition of the fact that there are no inherent weak links in these latter materials if they have been well made and properly inspected to eliminate any defectively manufactured pieces.

The inherent weaknesses of butt-welded and lap-welded seams arise from inability to achieve consistent, reliable bonding of the edges of the skelp via these "forge" welding processes. The degree of reliability of these materials was demonstrated by means of hundreds of burst tests conducted by National Tube Company.[2-9] The conclusions of National Tube work were:

- The average hoop stress for the bursting of butt-welded steel pipe was 73 percent of the ultimate tensile strength of the parent material.

- The average hoop stress for the bursting of lap-welded steel pipe was 92 percent of the ultimate tensile strength of the parent material.

- The vast majority of the failures of either kind of pipe initiated in the weld.

**Types of Defects**

The problem inherent with forged seams is the narrow temperature range over which adequate fusion can be effected. With today's cleaner steels, welding in a vacuum or inert environment might result in a high quality forged weld. Under such conditions the bonding surfaces would not be subjected to rapid oxidation and the sulfur would be too low to cause serious "hot shortness" a condition which leads to high temperature cracking. Unfortunately, the commonly used material for butt or lap-welded pipe was Bessemer steel which often contained relatively large amounts of sulfur and other harmful impurities. Furthermore, for economic reasons air was not excluded from the environment. As a result, the "welder" had to judge from the visual appearance of the heated tube materials in the furnace when the temperature was right for welding. Above 2460°F the oxide would remain molten and would be excluded from the bond zone by pressure exerted as the edges of the skelp were pressed together. If total exclusion was not effected, some or all of the mating surfaces would not bond but would instead entrap a weak, brittle nonmetallic layer. A lap-weld with entrapped oxide layers is shown in Figure 2-4.

Another defective condition was called "burning".[2-10] Typically, this condition occurred when the edges of the skelp were heated to too high a temperature and austenite grain-boundary sulfides formed. These layers essentially became cracks as shown in Figure 2-5 after the material had cooled. It is believed that the two extreme conditions, inadequate bonding due to oxide entrapment and burning accounted for the inability on the average of lap-welded joints to develop the full strength of the pipe. The fact that the seams of butt-welded pipe exhibited an even lower average percentage of the full strength of the pipe may be the result of the same factors and the fact that the bonded surface of a butt-welded seam is inherently smaller than that of a lap-welded seam.

## References

2-1.  Thomas, P. D., "Pipe-Manufacturing and Use", *Iron and Steel Engineer*, July 1957.

2-2.  Camp, J. M., and Francis, C. B., "The Making, Shaping, and Treating of Steel", Fourth Edition, 1925, Carnegie Steel Company.

2-3.  National Tube Company, General Catalog, 1935.

2-4.  Anon., "New Continuous Pipe Mill at Spang Chalfant", *Iron and Steel Engineer*, August 1938.

2-5.  Hess, L. J., "Some Operating Problems of Continuous Pipe Mills", *Iron and Steel Engineer*, September, 1941.

2-6.  Richardson, N. W., "The Manufacture of Butt-Welded Pipe",
      *Iron and Steel Engineer*, November, 1945.

2-7.  Rodder, W., "Developments in Continuous Butt Weld Pipe
      Mills", *Iron and Steel Engineer*, February, 1949.

2-8.  Anon., "Jones & Laughlin Puts Continuous Weld Pipe Mill on
      Production", *Blast Furnace and Steel Plant*, August, 1958.

2-9.  "Book of Standards", National Tube Company, Pittsburgh, PA,
      1913.

2-10. Hayes, V. L., and Rawlins, C. E., "The Evolution of
      Quality Line Pipe", 22nd Annual Pipe Line Conference, API,
      Dallas, TX, April 26-28, 1971.

WALL THICKNESS

O.D. SIDE

WIDTH OF SKELP

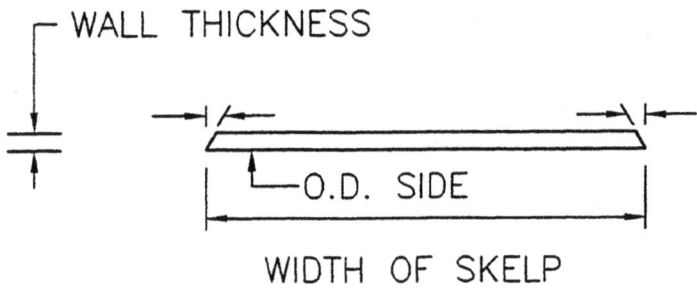

LENGTH DIRECTION

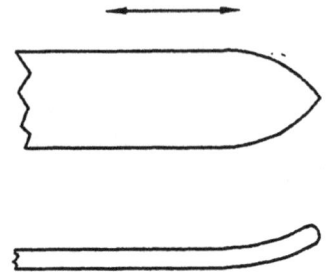

V-SHAPED END
FOR GRIPPING

a. SKELP FOR BUTT-WELDED PIPE

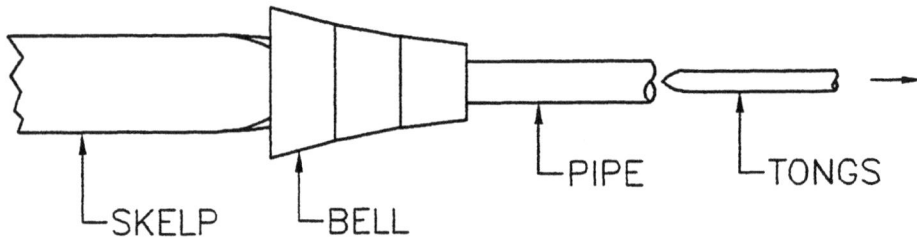

PIPE

TONGS

SKELP

BELL

b. THE WELDING BELL

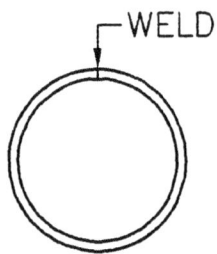

WELD

c. WELDED PIPE

d. SIZING ROLLS

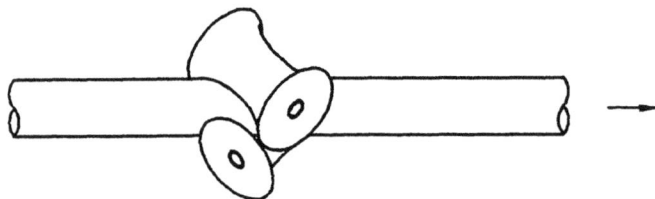

e. STRAIGHTENING ROLLERS

Figure 2-1.  Butt-Welded Pipe

SCARFED EDGES

WALL THICKNESS

OD SIDE

PIPE CIRCUMFERENCE

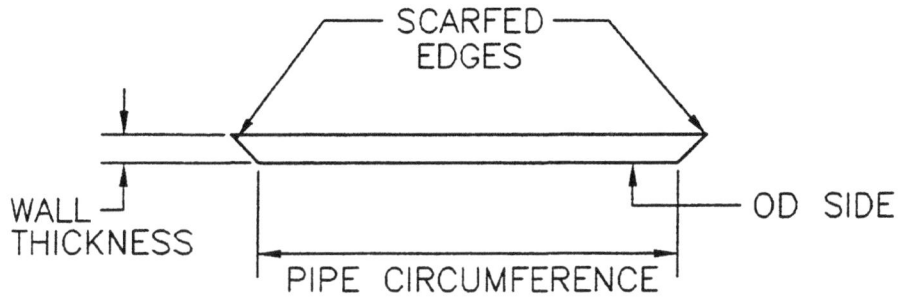

A. SKELP FOR LAP—WELDED PIPE

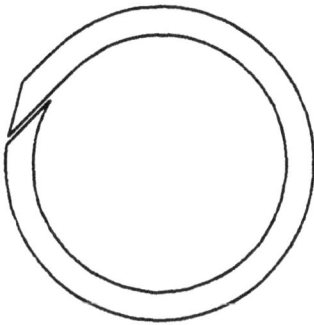

B. OVERLAP OF SCARFED EDGES AFTER BENDING

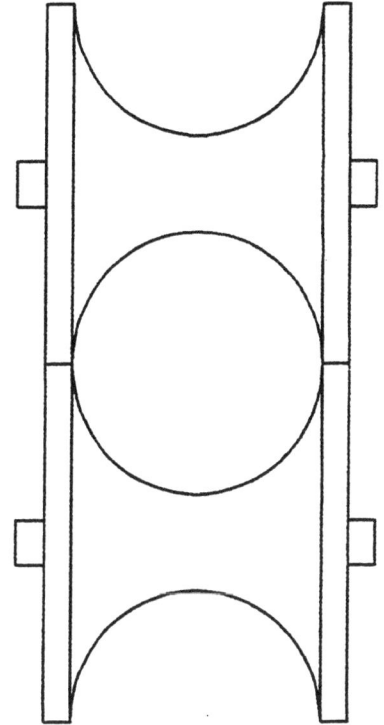

C. WELDING ROLLS

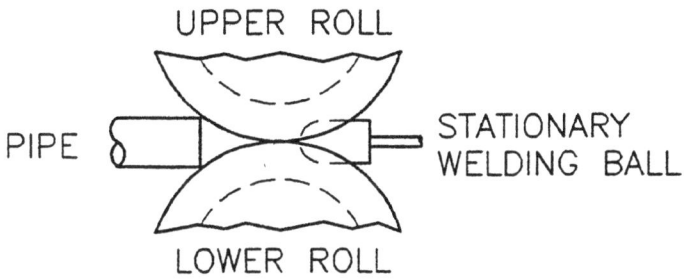

UPPER ROLL

PIPE

STATIONARY WELDING BALL

LOWER ROLL

D. WELDING TAKES PLACE AS PIPE PASSES BETWEEN SIZING ROLLS AND OVER WELDING BALL

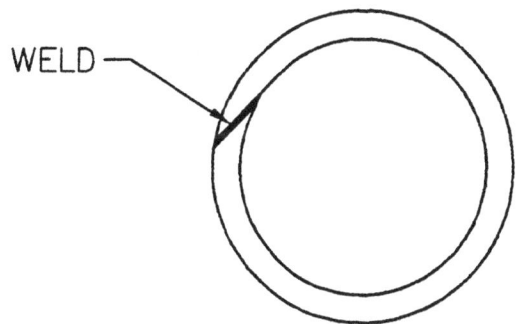

WELD

E. WELDED PIPE

Figure 2-2.   Lap-Welded Pipe

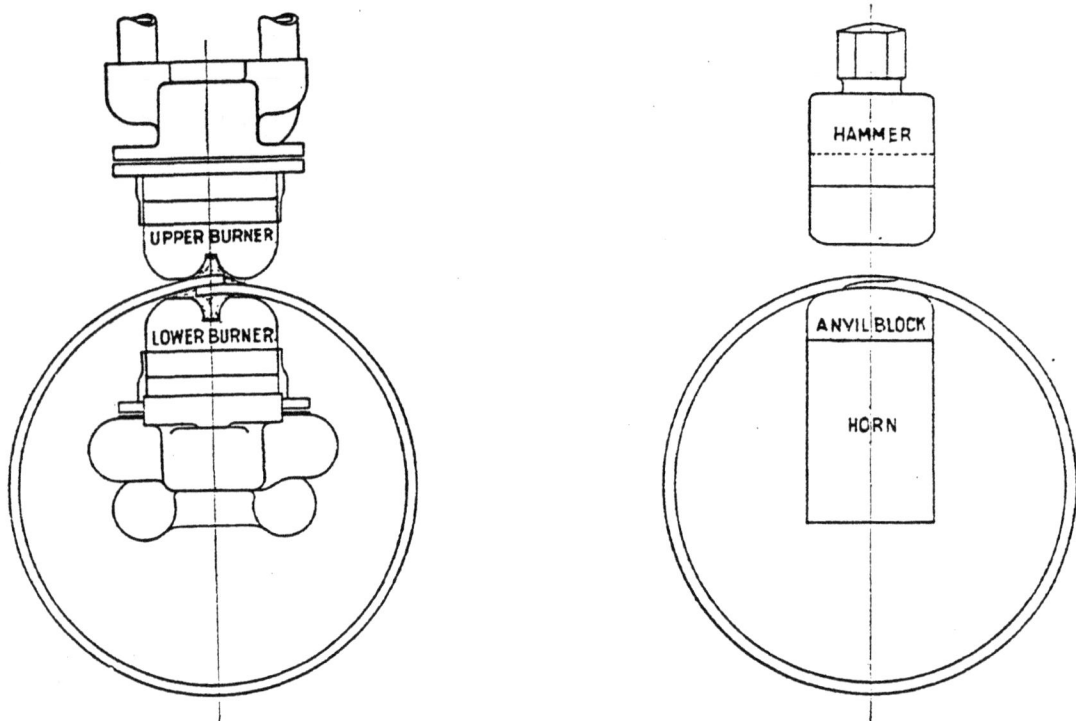

[Reprinted with permission from *The Making Shipping and Treating of Steel*, copyright 1985, Association of Iron and Steel Engineers]

Figure 2-3.   The Hammer Weld Process

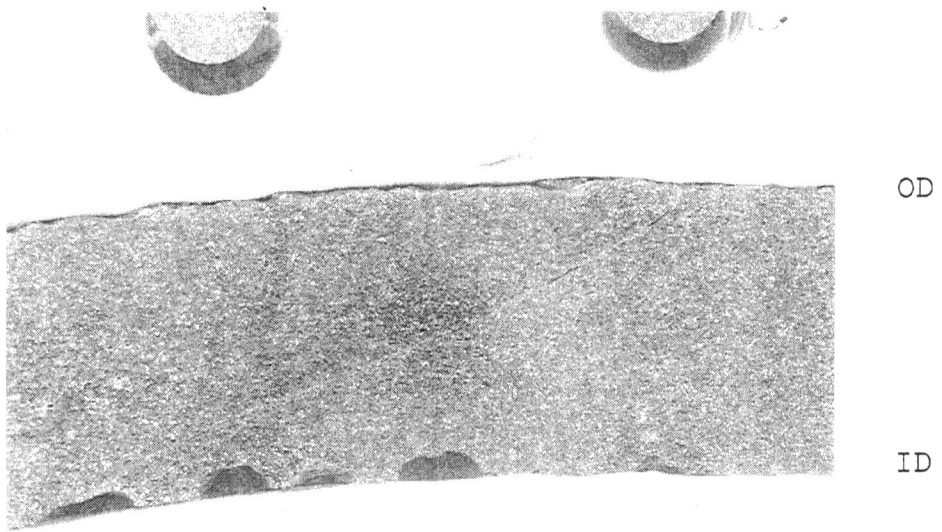

OD

ID

5X

(a)

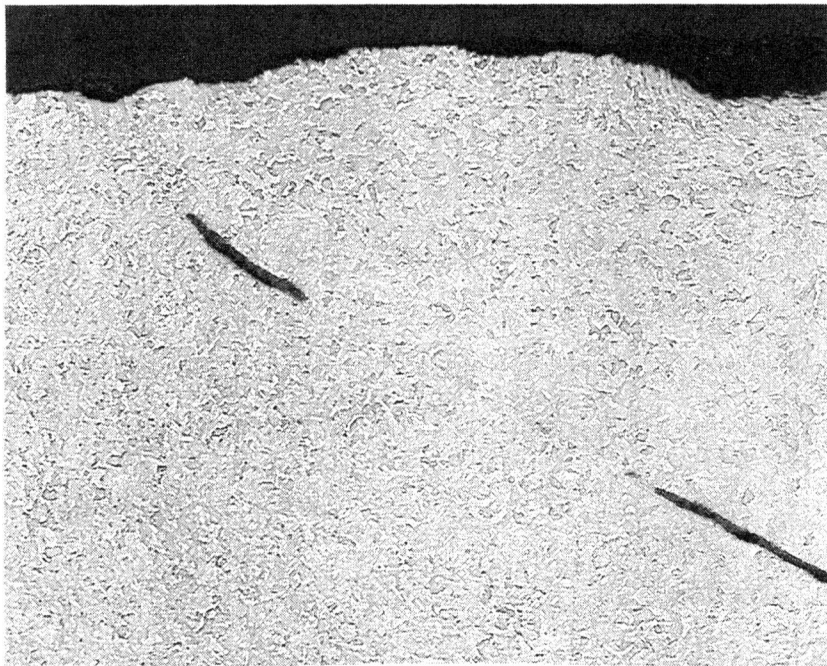

25X

(b) (Field reversed from (a) by system optics)

Figure 2-4. Lap Welded Seam Showing Entrapped Nonmetallic Material

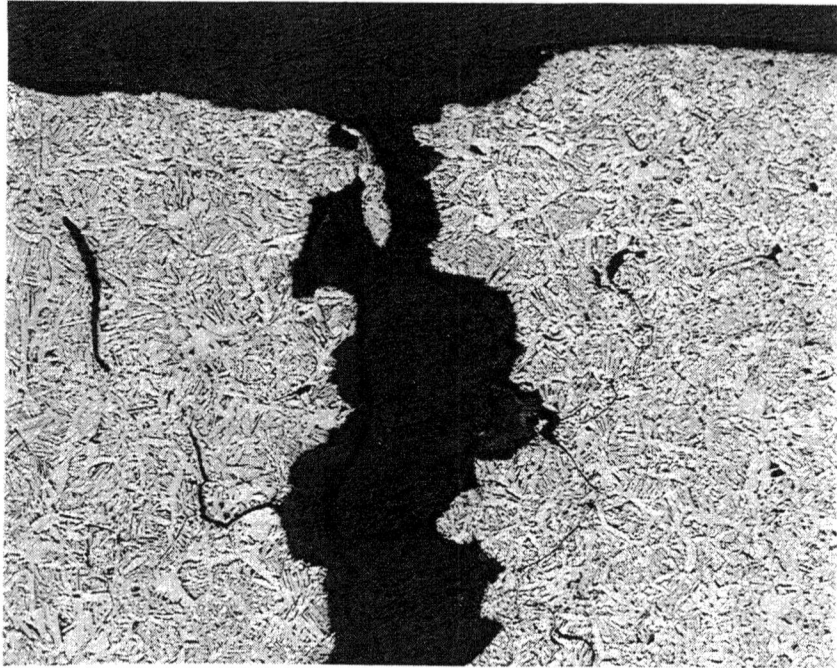

OD

25X

(a) Etched (2% Nital)

OD

(b) Unetched          25X

Figure 2-5.     Lap-Welded Pipe Exhibiting "Burned" Appearance Near
Bond Line

## 3.0 SEAMLESS LINE PIPE

The concept of seamless pipe is almost as old as that of butt-welded pipe. Whereas the latter began to be produced in practical quantities in 1824, the first successful methods for making seamless pipe appeared in 1840 and 1845.[3-1 to 3-3] One of these involved "piercing" a hot round billet held in a press followed by drawing or rolling to obtain the desired diameter, thickness, and length and the other involved "cupping" wherein a hot circular plate was pressed successively deeper by pairs of conical dies until a closed-end cylinder was formed. The development of the Mannesmann rotary piercing process in 1886 marked the beginning of the most-widely used (even today) process for producing large quantities of seamless line pipe.

Until the mid 1920s, lap-welded and butt-welded pipe made up the bulk of line pipe production. The rapid expansion of the oil and gas industry in the 1920s, however, led to increased demand for tubular goods including casing, tubing, and line pipe. These applications demanded a higher degree reliability and serviceability than had previously been required, and seamless pipe was better suited to these requirements than either lap-welded or butt-welded pipe. In the mid to late 1920s several U.S. manufacturers developed the capability to manufacture large-diameter double random lengths (~ 40 feet) of seamless line pipe and the age of modern line pipe and oil country tubular goods was born. Several process have been used and are used for making seamless pipe. These are described below beginning the most widely used type of mill, the plug mill.

### The Plug Mill

In the plug mill process, seamless pipe is made by first piercing a solid round billet which has been heated to a temperature in the range of 2200 to 2300°F.[3-4 to 3-8] The billet is

forced between two double-tapered rolls which lie in a plane
parallel to the axis of the billet but which revolve about skewed
axes as shown schematically in Figure 3-1. The space between the
rolls is called the "gorge". Upon entering the gorge the billet
encounters the stationary "piercing mandrel". The rolls impart a
helical motion to the billet and ovalize it to the point where it
begins to rupture in the center. The forward motion imparted by
the rolls pushes the billet with its rupturing center onto the
mandrel. The intent is to start the billet onto the mandrel at
the point of incipient rupture so that a smooth internal surface
is assured rather than letting an actual hole develop by
separation alone. The shaping of the tube is illustrated
schematically in Figure 3-2. The resulting hollow round tube
grows in length as the billet is consumed. This, essentially, is
the classic Mannesmann rotary piercing process. It is the first
step in almost all processes for making seamless pipe. But it is
only the first stage. Different processes utilize different
subsequent steps. The plug mill process for making seamless is
illustrated in Figure 3-3 and includes the following steps.

- A second piercing pass may be made without reheating.
  This is generally done for medium-to-large diameter
  pipe since one piercing pass cannot reduce the wall
  thickness sufficiently.

- Plug rolling (which gives the process its name) follows
  piercing. For larger sizes the rounds are usually
  reheated before plug rolling. As shown in Figure 3-4
  plug rolling continues the forming of the tube by
  slightly reducing the diameter and wall thickness while
  lengthening it. The rolls are shaped intentionally to
  slightly ovalize the round. The round is passed
  through the rolls at least twice, the second pass being
  made after it has been rotated 90 degrees. As the
  figure shows stripper rolls are engaged to push the
  round off of the plug.

- Rotary rolling (See Figure 3-5) is used for the largest sizes (16-inches and up). If used, it follows plug rolling. Rotary rolling is similar to piercing except that the radical shape of the rolls is designed especially to create a significant diameter enlargement without a great deal of change in wall thickness or length. Rotary rolling is usually done following a reheating of the round.

- After plug rolling (or after rotary rolling of the larger sizes) the rounds are subjected to reeling. The action of a reeling machine is similar to that of the rotary piercer except that the two skewed-axis rolls as shown in Figure 3-6 are almost cylindrical. The wall thickness reduction effected during reeling is slight. The intent is to round up the tube and improve the surfaces which may be left fairly rough from prior forming operations.

- Normally the last stage of seamless pipe hot working consists of sizing. Sizing is done by passing the reheated tube through a series of rolls which gradually bring it down to the final desired diameter and roundness.

After the pipes have cooled they are, of course, subject to end preparation, hydrostatic testing, and inspection in the same manner as welded line pipe.

Some additional processing may be done in special situations. While it is done rarely for line pipe, quenching and tempering is often done to produce the higher strength grades of API casing and tubing. In addition, in the past some seamless line pipe was cold expanded by a process unique to seamless pipe. Whereas welded pipe is cold expanded either by internal hydraulic pressure or in successive steps by an internal hydraulically-operated mechanical expander, the cold expansion of seamless pipe was done by pushing a lubricated plug clear through each length.

## The Mandrel Mill

The concept of the mandrel mill[3-8,3-9] may have been first used in Europe and a variation of it was used by Globe Steel Tubes Co. before 1950. In 1951 U.S. Steel installed a mandrel mill as Mill No. 4 at the Lorain works, calling it a "continuous" seamless pipe mill because it involved nine forming stands and could produce longer-than-normal lengths of tubes. As shown in Figure 3-7 the mandrel mill utilizes traditional rotary piercing as its first step in converting a solid billet to a round tube. Next a lubricated mandrel bar is inserted into the round, and the round and mandrel are started into the first stand. The series of rolls shown in Figure 3-7 (usually 7 to 9 stands) reduces the wall thickness of the tube to the desired value while significantly lengthening it. Generally, while the tube is rolled against the mandrel to reduce the thickness, the last passes are intended to ovalize the pipe slightly so that the process is finished with the tube having an inside diameter slightly larger than the outside diameter of the mandrel. Otherwise the tube could not be easily stripped from the mandrel. Once the process has been completed, the tube is stripped from the mandrel and the mandrel is cooled and relubricated for the next pass. The tube itself is usually then reheated and sent to a sizing mill or a stretch reduction mill. Mandrel mills are favored for producing smaller sizes such as 2⅜ to 4½-inch outside diameter pipe but are used to produce oil country tubular goods in sizes up to 9⅞-inch outside diameter and line pipe in sizes up to 8⅝-inch outside diameter.

## Assel and Transval Mills

Assel and Transval mills are generally used in the manufacture of heavy-wall tubing in sizes up to 8⅞-inch outside diameter with D/t ratios ranging from to 10:1 to 15:1[3-8]. However, D/t ratios as high as 22:1[3-10] and as low as 3½:1[3-11] are made. Obviously, these types of materials would not often be used as line pipe, but they could be used occasionally for deep-water, offshore applications. The processing steps in an Assel mill are shown in Figure 3-8. As with all processes the billets are piecered by traditional rotary piercing. The Assel "elongator" is a three-roll arrangement as shown in Figure 3-9. An Assel elongator "cross" rolls the tube unlike a plug mill or mandrel mill which provides only longitudinal rolling. Note that the single pass through these rolls creates a radical step-down in wall thickness. A internal mandrel is utilized to maintain the inside diameter. After being formed in the Assel mill, the tube is reheated and passed to a hot reduction mill and a rotary sizing mill as shown in Figure 3-8.

One reason an Assel mill is generally not used to make light-wall tubes is that the three-roll elongator has a tendency to cause severe distortions at the beginning and ends of the tube. A Transval Assel mill is equipped with a quick-change orientation feature that allows, stepped-down and stepped-up thicker walls at the beginning and ends of each tube. Therefore, it can be used to roll somewhat lighter-wall materials than a standard Assel mill.

## The Diescher Mill

A Diescher elongator is shown in Figure 3-10. This type of elongator cross rolls a previously rotary-pierced tube between two tapered rolls against a freely rotating internal mandrel while two powered guide disks in the plane 90 degrees to the cross rolls help guide and shape the tube. A version of a Diescher mill called the "ACCU-ROLL" elongator is described in Reference 3-11 The ACCU-ROLL concept is illustrated in Figure 3-10. It utilizes a restrained rather than a full-floating mandrel. After being elongated in this manner the tube passes through a final sizing mill step. Diescher-type mills feature especially well controlled wall thickness and inside diameter and outside diameter surface finishes.

## Other Processes

### The Stiefel Disk

A Stiefel disk is a type of billet piercer that was developed in 1895 and used early in this century at National Tube company's Ellwood Works[3-12]. Eventually it was replaced by the Mannessmann-type piercer because it was not well-suited for making large outside diameter tubes. A Stiefel piercer is illustrated in Figure 3-12. In principle it did the same thing as a Mannessmann piercer.

### The Pilger Mill

Pilgering is a rolling operation developed by Mannessmann shortly after their development of rotary piercing.[3-1,3-2] For rolling in a Pilger mill a solid bar mandrel is inserted into a rotary-pierced tube and the tube with mandrel is pushed into the Pilger throat.[3-13] Each roll in the pair of Pilger rolls shown in Figure 3-13 has a varying width throat and an abrupt transition. Since the rolls rotate counter to the

direction that would draw the tube into the throat and since part
of the throat is too narrow for the tube to pass, the tube is
seized and forged by the impact at some point.  The tube is
forced back against a hydraulic shock absorber in the mandrel,
rotated 90 degrees, and pushed again into the Pilger throat.
This action is repeated as often as needed until the entire tube
is reduced to a wall thickness that will pass the Pilger throat.
The Pilgered tube is then passed onto a reeling mill and a sizing
mill.

At least five North American manufacturers used Pilger
mills at one time:  National Tube Company (Ellwood Works and
Christy Park Works), Youngstown Sheet and Tube Company (Campbell
Works), Phoenix Steel Corporation (Phoenixvile, PA), Pittsburgh
Steel Company (Allenwood Plant), and Tubos de Acero de Mexico,
SA.  The Pilger mill at Ellwood was replaced in 1906 when
National Tube developed the plug mill process.[3-3]  The Pilger
mill at Youngstown was replaced by a plug mill in 1938.[3-3,3-4]
The histories of the Pilger mills at Pittsburgh Steel, Tubos de
Acero, and National Tube Company's Christy Park Works are not
known.  Up to at least the late 1980s the Pilger mill at Phoenix
Steel was still operating.  They apparently referred to the
process as "rotary forging".  Pilger mills were relatively slow
compared to plug mills due to the numerous passes required.  Also
they were harder to maintain.  Probably they were abandoned for
economic reasons.

**Press Piercing**

A new type of piercing called press piercing[3-14] which
sounds much like one of the original techniques[3-1] is being used

to make 3½- through 9⅝-inch tubes at USX Corporation in Fairfield, AL.  Following the press piercing is a conventional mandrel mill and a stretch reduction mill.

## Quality Factors

While the problems sometimes associated with welded seams are avoided, seamless pipe is not without its own unique set of potential problems.  The potential problems fall into the following categories.

- Material quality
- Processing variables
- Imperfections and defects introduced by forming
- Mechanical properties

### Material Quality

Very early on it was discovered that poor quality material (i.e., dirty or segregated steel) produced severe defects in seamless pipe owing to the variety of forces induced during forming.  Whereas in welded pipe made from plate or coiled skelp the flat rolling tends to conceal or neutralize some types of nonmetallic inclusions and laminations, the multi-directional forging operations during the making of seamless pipe often magnify the effects of these anomalies.  Voids or nonmetallic inclusions can foul up piercing to the point where the mandrel and/or the pipe is ruined.  Laminations, instead of ending up on a flat plane at mid wall as in a plate, tend to slope through the thickness over a portion of the circumference, weakening the hoop-stress carrying capacity of the tube.  It was quickly learned in the early days that Bessemer steels were not suitable for seamless pipe and that hot-topped (cropped) ingots of fully-deoxidized steels were needed as feed-stock for seamless tube mills. [3-2,3-4,3-15]

Ingot practice was found to be important. Blow holes, cracks and scale had to be eliminated. Whether the billets were made as cast billets, blooms, or rolled rounds it was necessary to prepare them for making seamless pipe. Large rounds or ingots were often flame-scarfed over their entire surface to remove surface imperfections. [3-15] Some mills "peeled" billets, a process involving machining away ¼-inch or more of the surface in a large lathe. [3-13]

Continuous casting has led to better quality billets. [3-16 to 3-18] By eliminating cropping of ingots and the need for peeling, continuous casting is said to have increased the yield to 96 percent as opposed to 75 percent for rounds produced from ingots. [3-16] Also, the newer, cleaner steels made in basic oxygen furnaces would be expected to reduce the problems associated with dirty material.

**Processing Variables**

The most important processing variable according to several authors [3-4,3-9,3-17] is the uniformity of heating. Non uniform heating leads to eccentricity during rotary piercing because the mandrel tends to move toward the hotter material. This condition cannot be corrected by subsequent processing. The authors of Reference 3-17 claim that if billet temperature variations are held to ±20°F the eccentricity due to nonuniform heating can be reduced to 2.5 percent. In contrast, if a 100°F variation is permitted the eccentricity can range from 5 to 9 percent. Uniform heat requires the observance of minimum soaking times in the rotary furnaces.

## Imperfections and Defects

Imperfection and defects arise both as carryover from billet defects and from forming irregularities. Scabs, blisters, slivers, seams, laps, laminations, pits, roll-ins, roll-marks and plug scores are the typical imperfections associated with seamless pipe. Seams arise from crevices carried over from billets or ingots that are pressed closed but cannot fuse to sound metal because of scale or dirt (See Figure 3-14). Laps arise during tube-forming operations when a portion of the material is folded over onto the tube without fusing with it (See Figure 3-15). Scabs and blisters are a thin veneer-like layer of steel attached to the surface but separated by a nonmetallic layer (See Figures 3-16 and 3-17). They usually arise from ingot or billet defects and may peel off in service. Laminations are a common result of inadequately cropped ingots. Roll-ins and pits arise from material trapped between the mandrel and the tube. In a roll-in the foreign material remains trapped (see Figure 3-18); a pit is all that remains if the foreign material falls out (see Figure 3-19). Slivers may be caused by trapped or rolled-in scale. They are usually attached by sound metal at one end, and when present on the outside diameter surface, they have the annoying habit of interfering with the proper application of coatings because they tend to be "popped-up" by grit blasting.

The many processing steps which involve twisting, upsetting, and abrading the hot material account for the surface roughness of seamless pipe and occasionally lead to defects such as laps, roll-ins, pits, roll marks and seams. Laps result from upset material which is folded over on the tube's surface but does not fuse. Not surprisingly, these as well as seams are frequently helically oriented. However, double piercing, where the direction of rotation of the second piercing is opposite from that of the first, may leave laps and seams oriented nearly longitudinally.  Plug scores and roll-marks usually tend to be long, longitudinally-oriented grooves.  Roll-ins and pits result from trapped foreign materials.

The key to minimizing defects in seamless pipe is process control.  By starting with relatively clean billets or rounds and by careful attention to mill tolerances, the manufacturer can prevent many of the harmful imperfections described previously.  Additionally, many types of defects can be eliminated by the final inspection.  As indicated in Reference 3-19, magnetic flux leakage techniques are available to scan both outside diameter and inside diameter surfaces.  These techniques can be expected to find the most potentially serious types of defects such as laps and seams.

Seamless pipe on occasion has been found to exhibit transit fatigue.  For a discussion of transit fatigue see Page 6-5.

## Mechanical Properties

Historically unless heat-treated, seamless pipe is finished hot and allowed to cool relatively slowly.  This minimizes the manufacturer's ability to limit grain growth.  As such the actions that can be taken to effect certain desirable properties such as high strength, low fracture propagation transition temperature, and high notch toughness are much more limited than is the case with welded pipe.  The latter can be thermomechanically treated to obtain both high-strength and high toughness as well as a low transition temperature.  For this reason as-rolled seamless line pipe has seldom been available in grades above X60 and was often available only in limited wall thicknesses in Grades X52 and X60.  This situation may be changing. Seamless manufacturers are currently experimenting with thermomechanical treatment. If perfected, these techniques could lead to higher grades of seamless pipe with improved toughness properties. It is, of course, possible to heat treat seamless pipe by quenching and tempering, or by normalizing to obtain the higher grades, but this may put seamless pipe at a cost disadvantage relative to welded pipe. These limitations should be kept in mind by the purchaser especially if high toughness or low transition temperature is vital to the use of the pipe.

## References

3-1.    "The Making, Shaping and Treating of Steel", Eighth Edition, 1964, United States Steel Corporation.

3-2.    Hamilton, N., "Some Observations on Seamless Tube Making", *Iron and Steel Engineer*, March 1946, pp 55-61.

3-3.    Ohlson, C., "Review of Developments in Seamless Tube Manufacture", *Iron and Steel Engineer*, August 1939, pp 15-18.

3-4.    Ess, T. J., "Youngstown Sheet and Tube Company Expands Seamless Facilities", *Iron and Steel Engineer*, April 1939, pp Y-1, Y-11.

3-5.    Anon., "Seamless Tubes Made in Mill of Latest Design at Plant of Pittsburgh Steel Company", *Blast Furnace and Steel Plant*, April 1945, pp 464-468.

3-6.    Anon., "Automatic Seamless Tube Mill Now in Production", *Blast Furnace and Steel Plant*, October 1957, pp 1145-1146.

3-7.    Crawford C. C., "Seamless Pipe for the Oil Fields", *Iron and Steel Engineer*, March 1956, pp 96-99.

3-8.    Schuetz, G. W., "Current Trends in Seamless Tube Mill Design", *Iron and Steel Engineer*, September 1976, pp 47-57.

3-9.        Young, J. L., "The Continuous Seamless Pipe Mill", *Iron and Steel Engineer*, April 1951, pp 53-61.

3-10.       Pozsgay, Albert, "ACCU-ROLL: A New Type of Seamless Tube Mill", *Iron and Steel Engineer*, June 1986, pp 36-42.

3-11.       Laird, W. W., Palko, S. J., and Marcouiller, R. J., "Ambridge—A Fully Computerized Seamless Tube Mill", *Iron and Steel Engineer*, January 1980, pp 69-73.

3-12.       Camp, J. M., and Francis, C. B., "The Making, Shaping, and Treating of Steel", Fourth Edition, 1925, Carnegie Steel Company.

3-13.       Oil Country Tubular Goods, Line Pipe, and Standard Pipe Catalog, No. 320L, Pittsburgh Steel Company, May 15, 1944.

3-14.       Subek, T., "Water Treatment System for the Seamless Pipe Mill of USS Fairfield Works", *Iron and Steel Engineer*, July 1988, pp 26-28.

3-15.       Petrosksas, J. A., "Top Pouring Ingots for Direct Rolling Seamless Tubes", *Iron and Steel Engineer*, October 1964, pp 111-115.

3-16.       Spor, R. W., "Modification of the J&L Aliquippa Billet Caster to a Rounds Caster", *Iron and Steel Engineer*. December 1982, pp 25-28.

3-17.       Code, J. B., "Algoma's New No. 2 Seamless Mill: The Process and Its Capability", *Iron and Steel Engineer*, October 1987, pp 26-30.

3-18.       Rau, R. J., "LTV Steel Campbell Works: A World Class 16-Inch Seamless Mill", *Iron and Steel Engineer*, December 1984, pp 35-40.

3-19.       Monks, E. S., "An Inspection Method Improvement Program for Seamless Product at Republic's Chicago District", *Iron and Steel Engineer*, September 1983, pp 17-21.

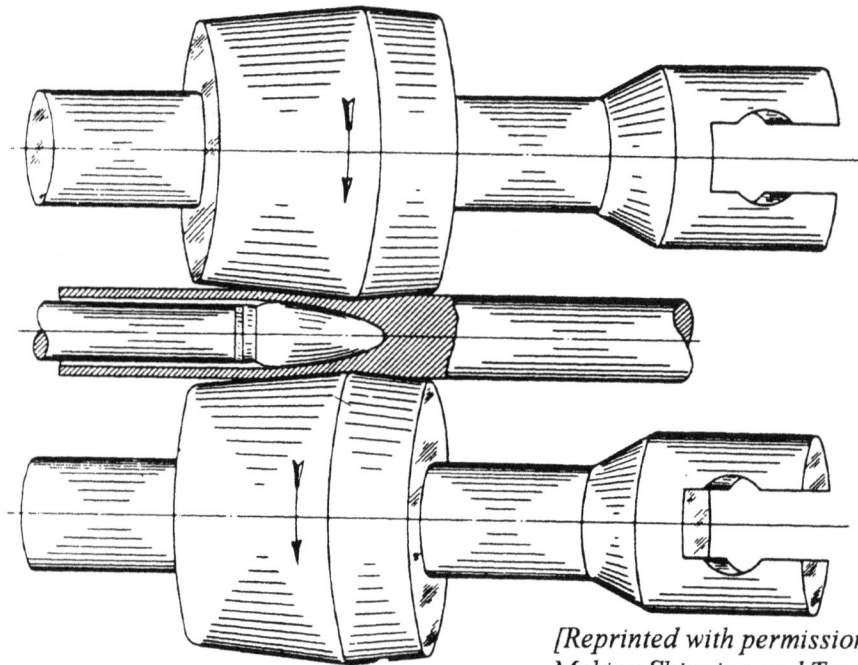

Figure 3-1. Mannesmann rotary piercing

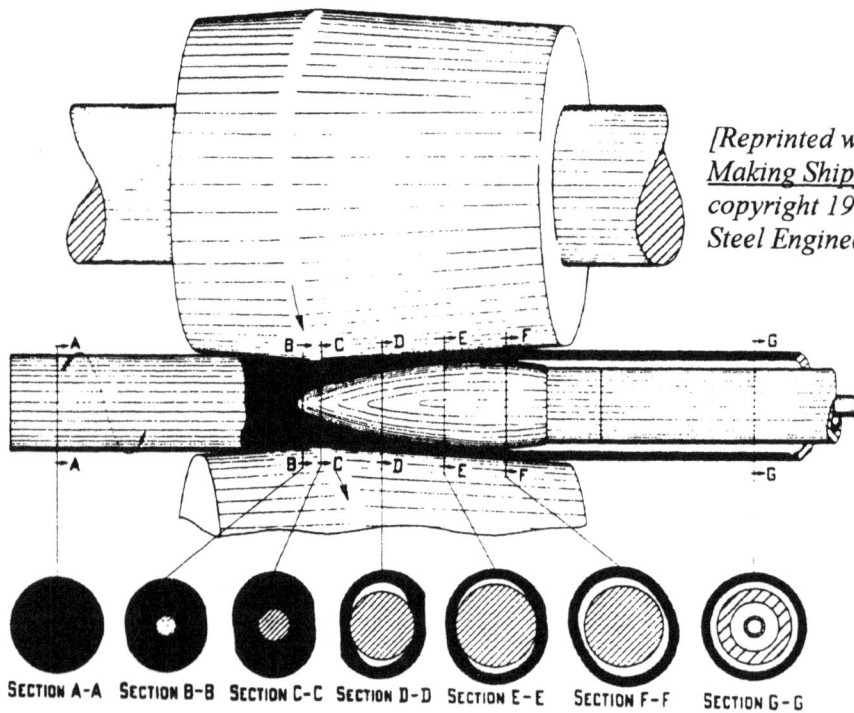

SECTION A-A  SECTION B-B  SECTION C-C  SECTION D-D  SECTION E-E  SECTION F-F  SECTION G-G

Figure 3-2. Forming action during rotary piercing.

ROTARY HEARTH FURNACE     NO. 1     NO. 2

SOLID ROUND BILLET

PIERCING MILLS

REHEAT FURNACE     PLUG MILL     REELER     REELER

SIZING MILL

REHEAT FURNACE     STRETCH REDUCING MILL     STRAIGHTENER

TO FINISHING FLOOR

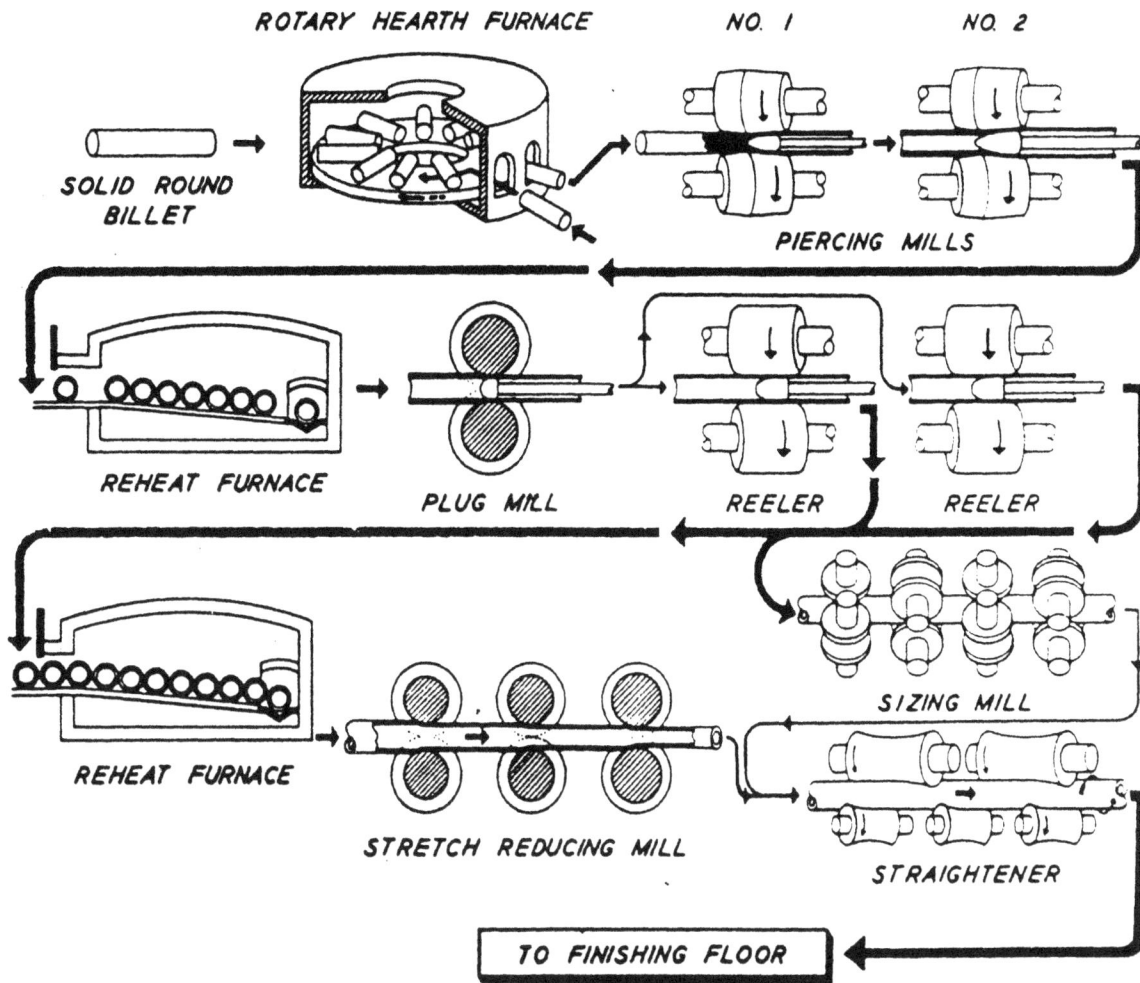

Figure 3-3.    The plug mill process.

POSITION OF
ROLLING MILL ROLL
WHEN STRIPPING

℄ ROLLING MILL

℄ STRIPPER ROLLS

ROLLING MILL
ROLL IN
ROLLING POSITION

STRIPPER ROLLS IN
STRIPPING POSITION

POSITION OF STRIPPER
ROLLS WHEN ROLLING

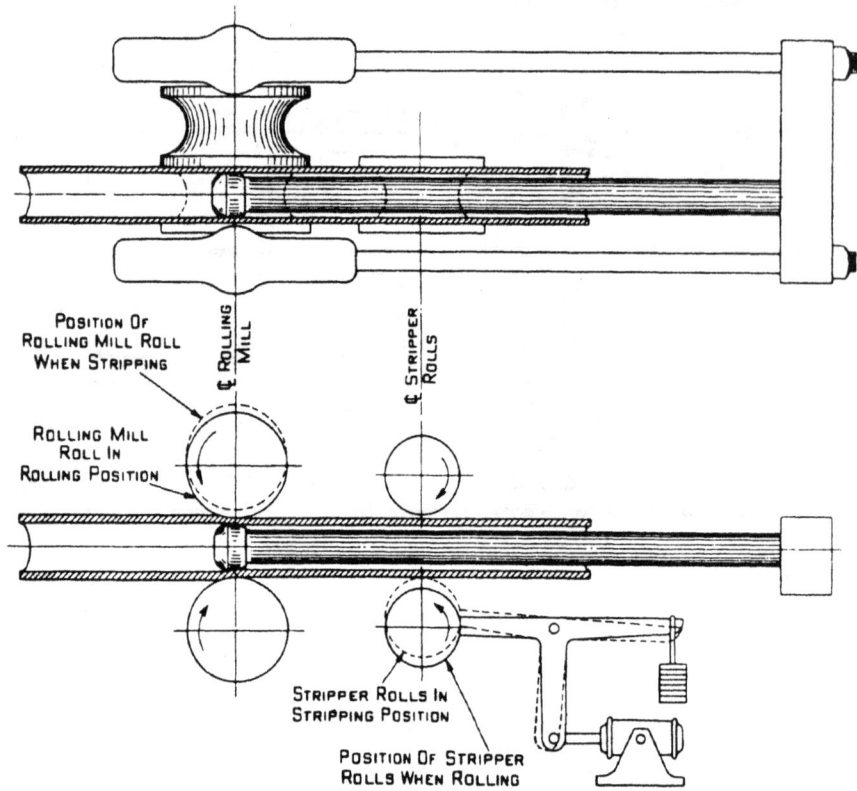

Figure 3-4. Plug rolling.

*[Reprinted with permission from The Making Shipping and Treating of Steel, copyright 1985, Association of Iron and Steel Engineers]*

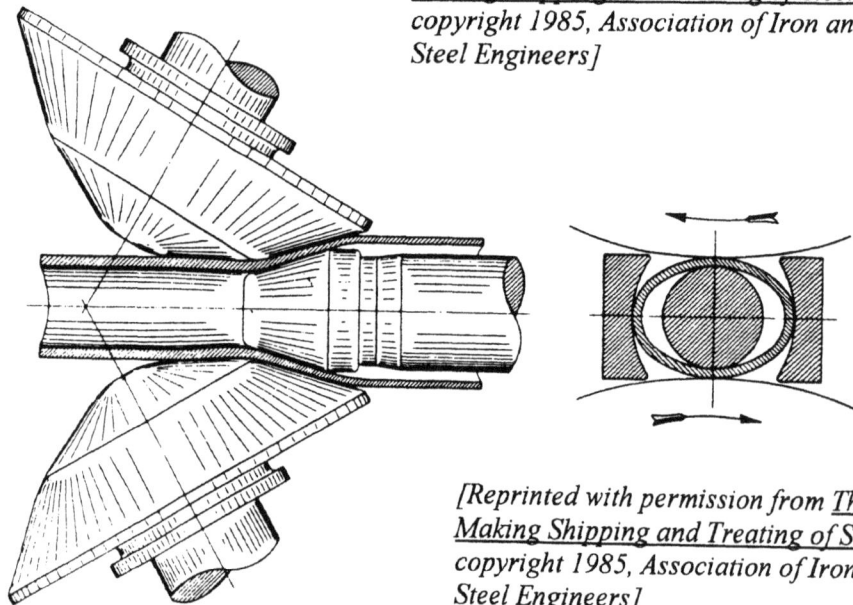

*[Reprinted with permission from The Making Shipping and Treating of Steel, copyright 1985, Association of Iron and Steel Engineers]*

Figure 3-5. Rotary rolling.

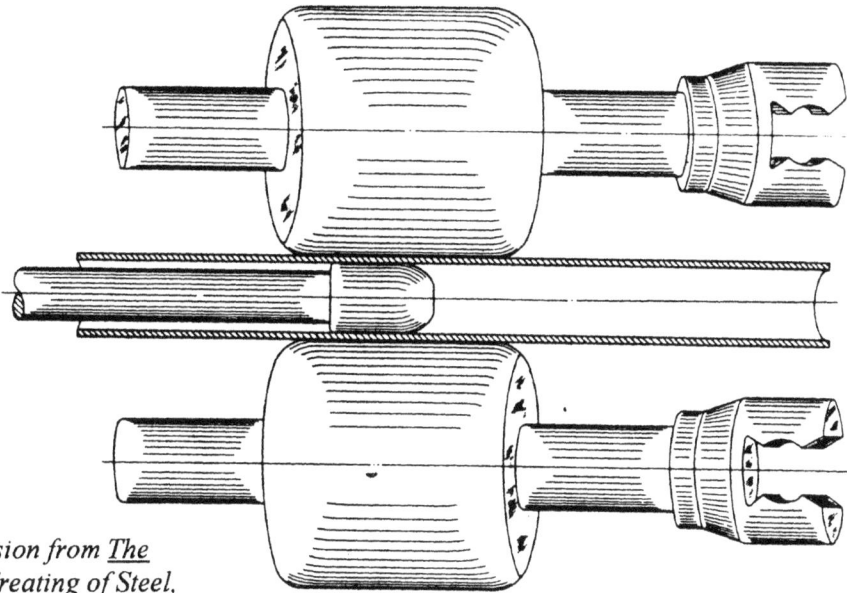

[Reprinted with permission from *The Making Shipping and Treating of Steel*, copyright 1985, Association of Iron and Steel Engineers]

Figure 3-6.  Reeling.

ROTARY HEARTH FURNACE

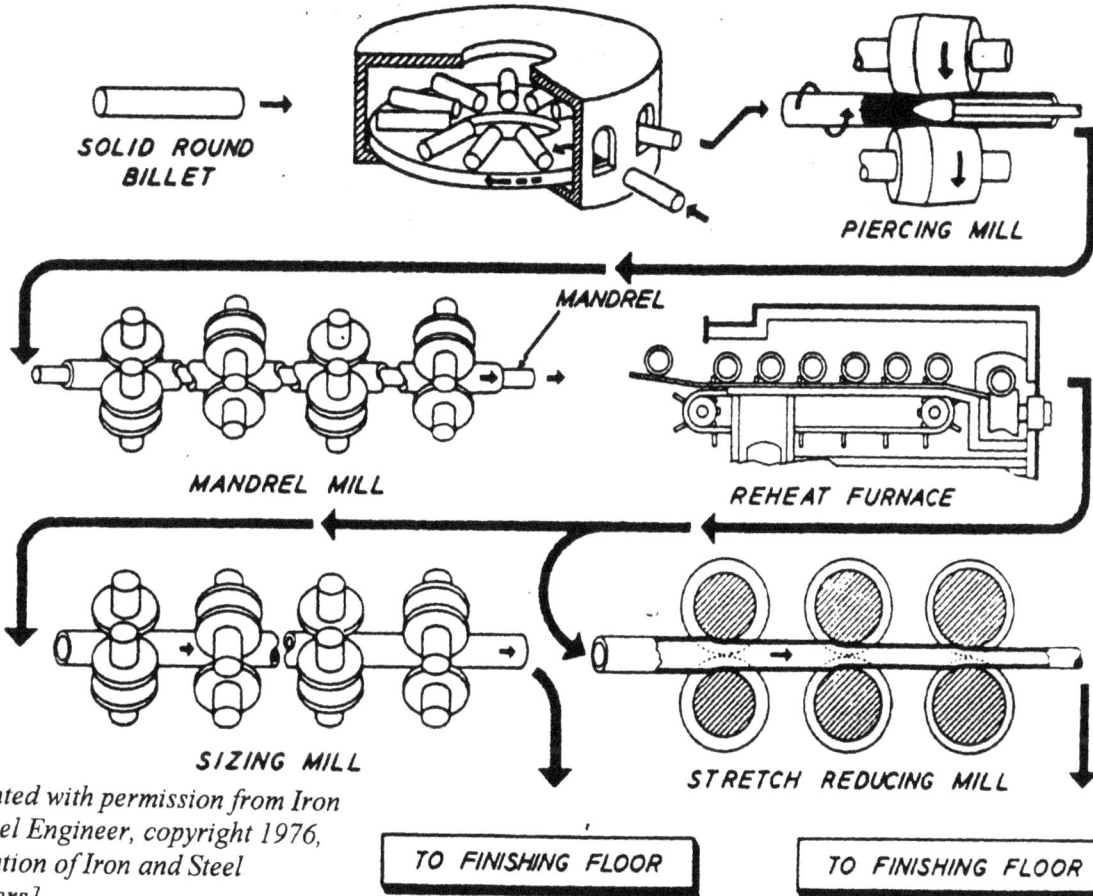

SOLID ROUND BILLET

PIERCING MILL

MANDREL

MANDREL MILL

REHEAT FURNACE

SIZING MILL

STRETCH REDUCING MILL

[Reprinted with permission from Iron and Steel Engineer, copyright 1976, Association of Iron and Steel Engineers]

TO FINISHING FLOOR

TO FINISHING FLOOR

Figure 3-7.  The mandrel mill.

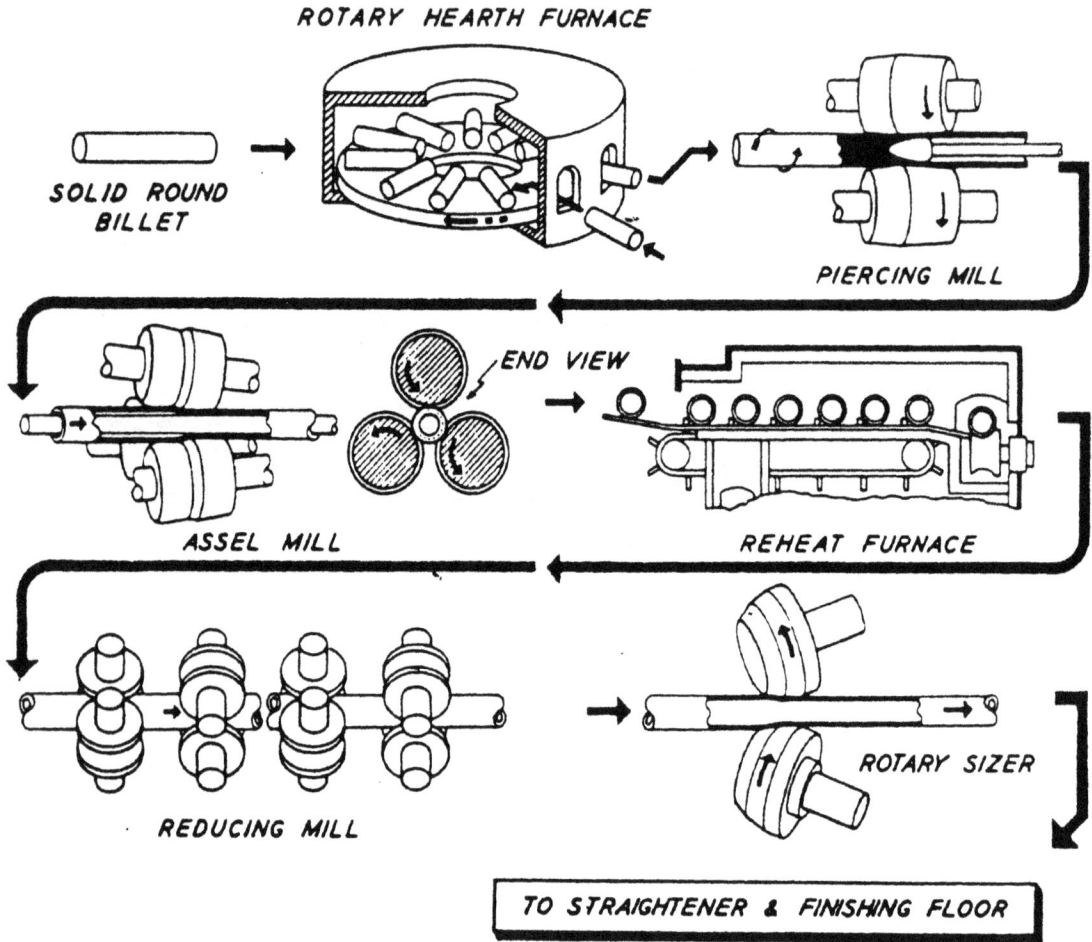

Figure 3-8.    The Assel mill.

*[Reprinted with permission from Iron and Steel Engineer, copyright 1976, Association of Iron and Steel Engineers]*

MANDREL

REDUCED OUTSIDE
DIAMETER OF TUBE

HEIGHT
OF HUMP

ROLLS

OUTSIDE DIAMETER
OF TUBE

MANDREL

SECTION A-A

TOP ROLL

A

A

BOTTOM
ROLL

BOTTOM
ROLL

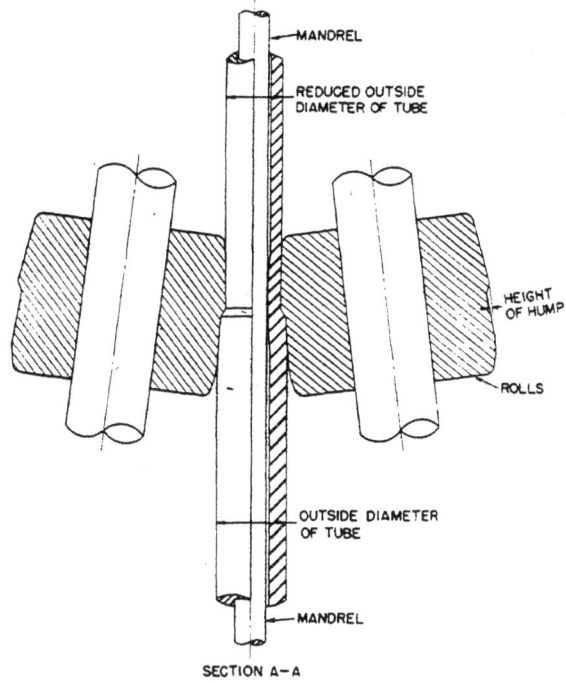

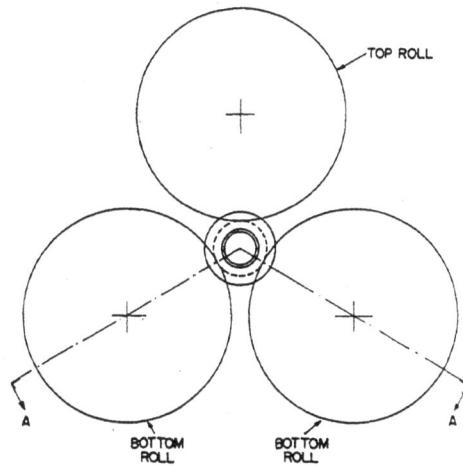

Figure 3-9.   The Assel mill rolls.

*[Reprinted with permission from The*
*Making Shipping and Treating of Steel,*
*copyright 1985, Association of Iron and*
*Steel Engineers]*

PLAN VIEW OF DIESCHER MILL
SHOWING ELEVATION AND SECTIONS

CROSS-SECTION OF DIESCHER ELONGATOR SHOWING
RELATIONSHIP OF GUIDES, DISCS AND ROLLS

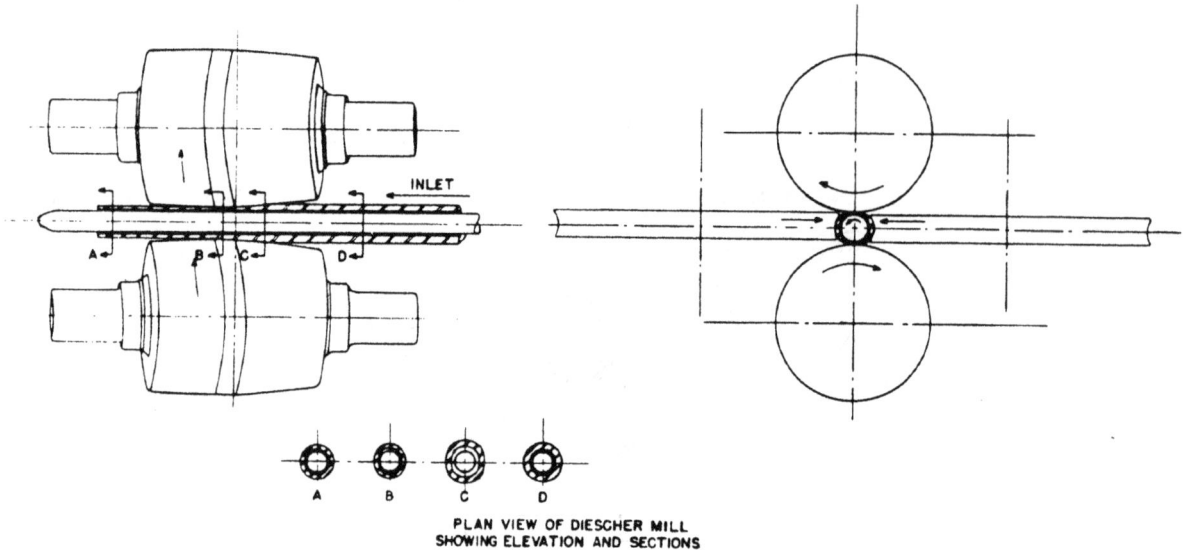

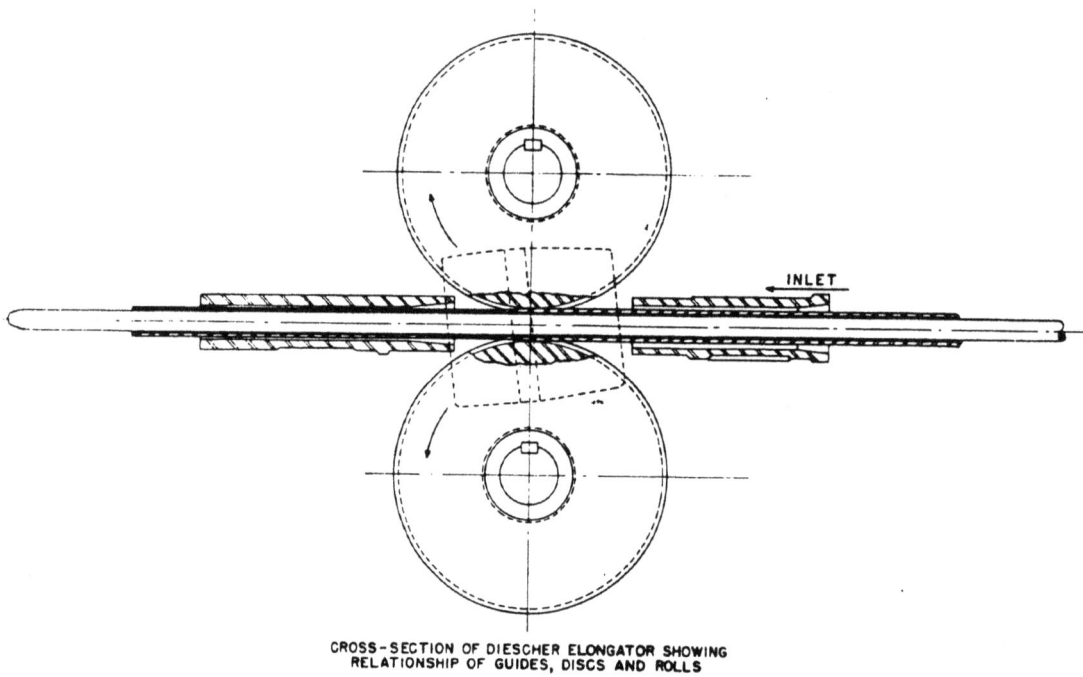

Figure 3-10.   The Diescher elongator.

*[Reprinted with permission from The
Making Shipping and Treating of Steel,
copyright 1985, Association of Iron and
Steel Engineers]*

Roll

Mandrel Bar

Disk Guide

Mandrel Bar

Roll

Disk Guide

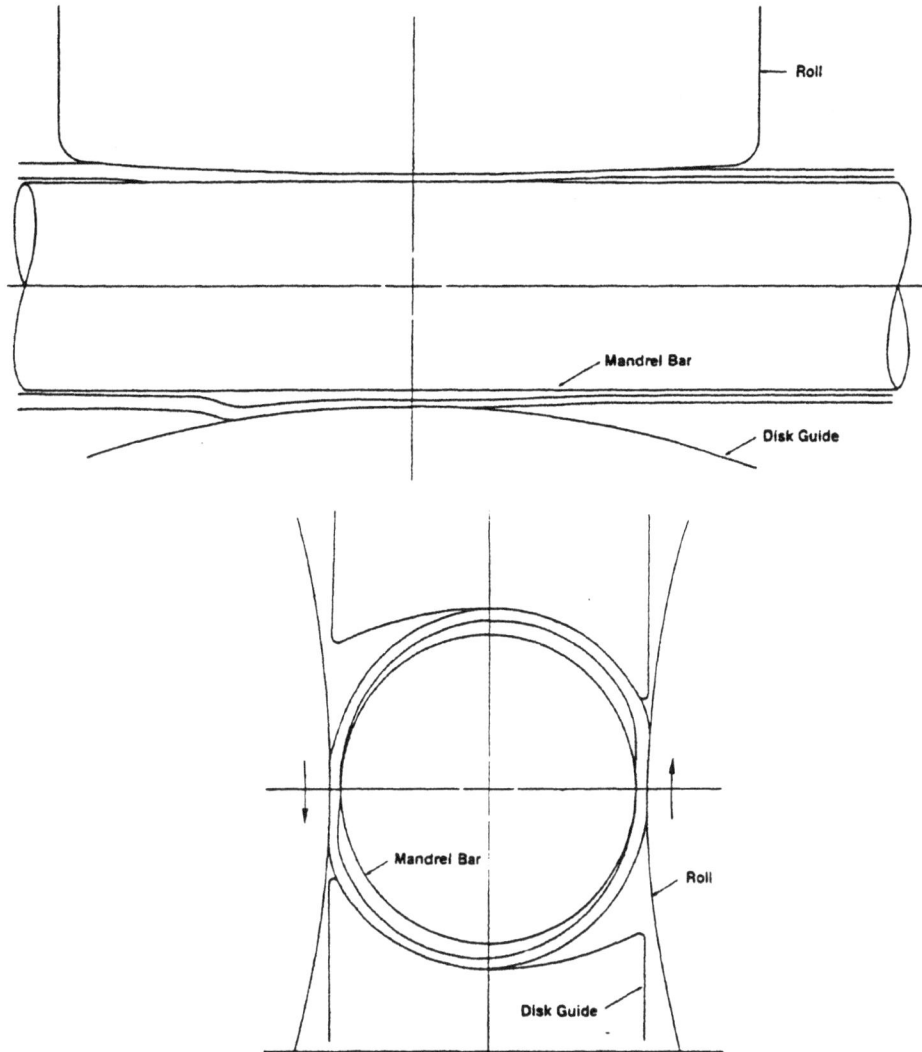

Figure 3-11. An "ACCU-ROLL" elongator.

*[Reprinted with permission from Iron
and Steel Engineer, copyright 1986,
Association of Iron and Steel
Engineers]*

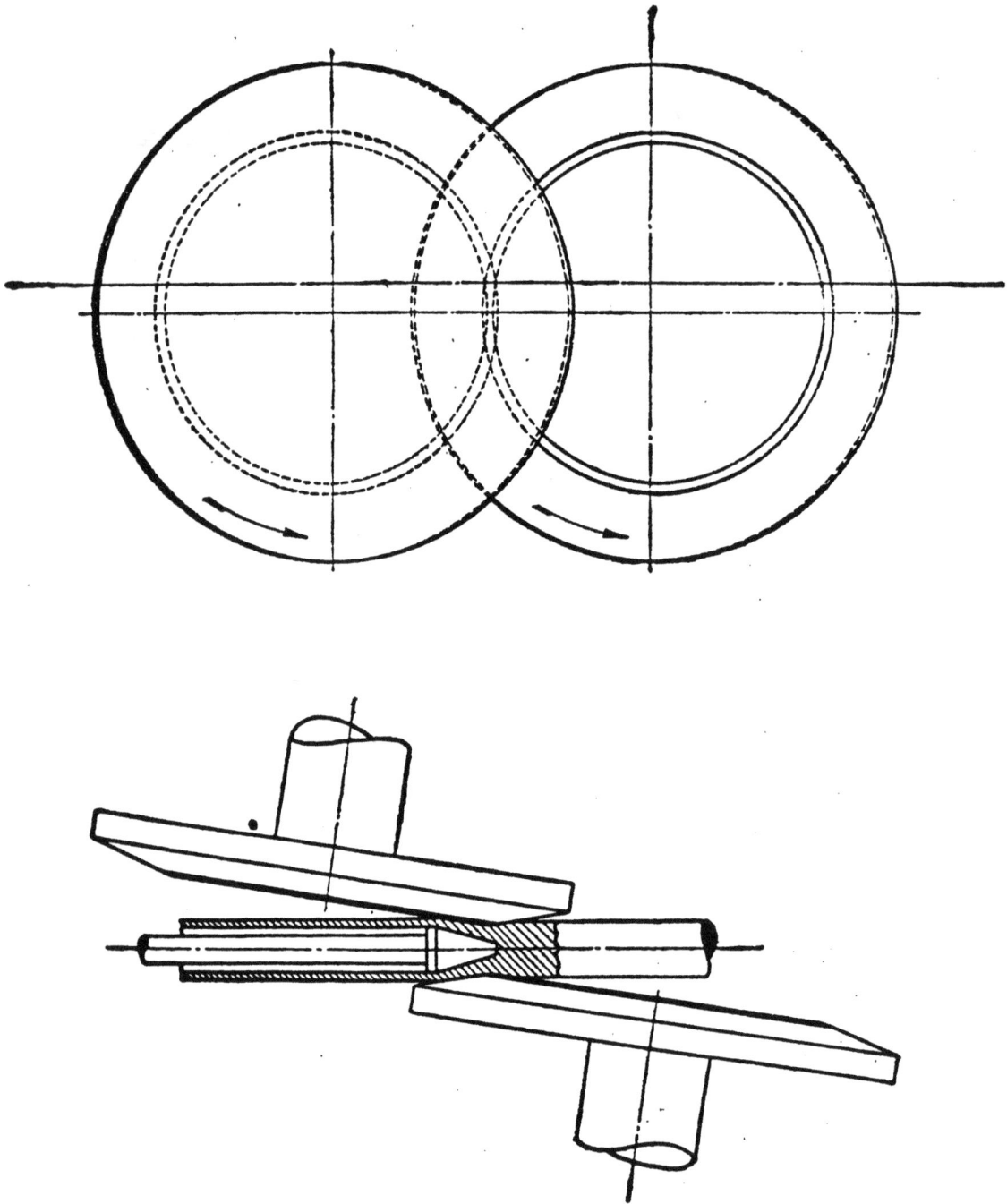

Figure 3-12.   The Stiefel Piercer.

*[Reprinted with permission from The Making Shipping and Treating of Steel, copyright 1985, Association of Iron and Steel Engineers]*

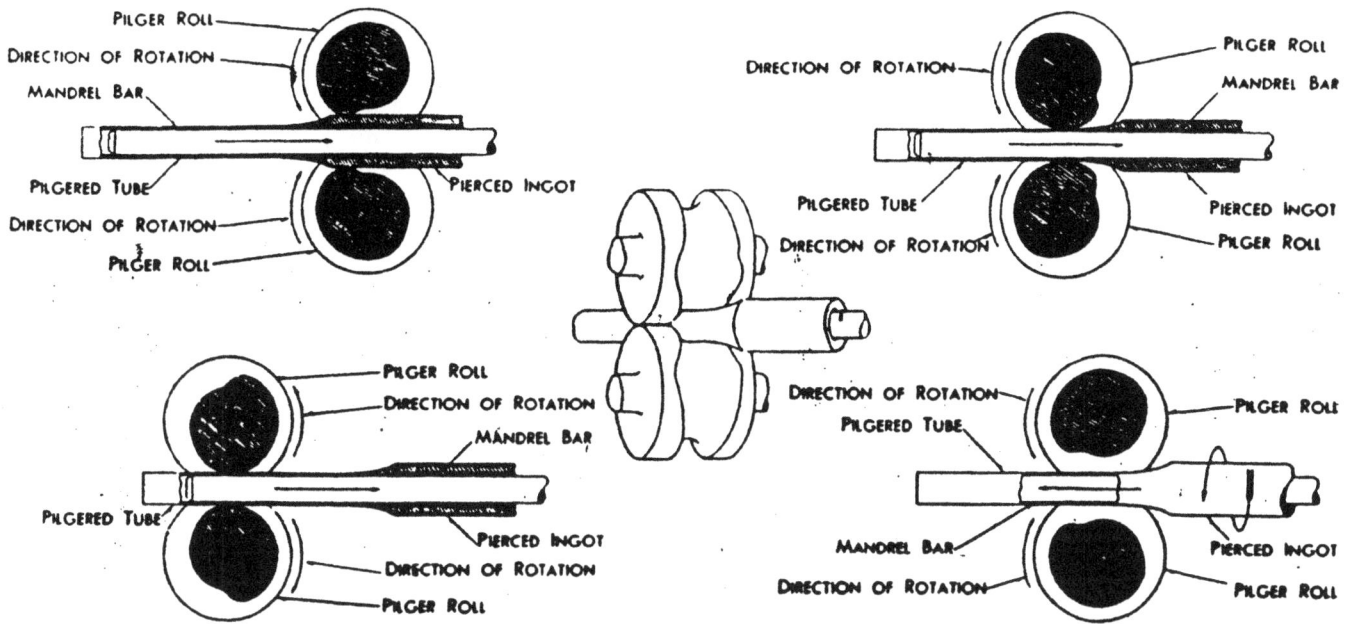

Figure 3-13.   Pilger rolls.

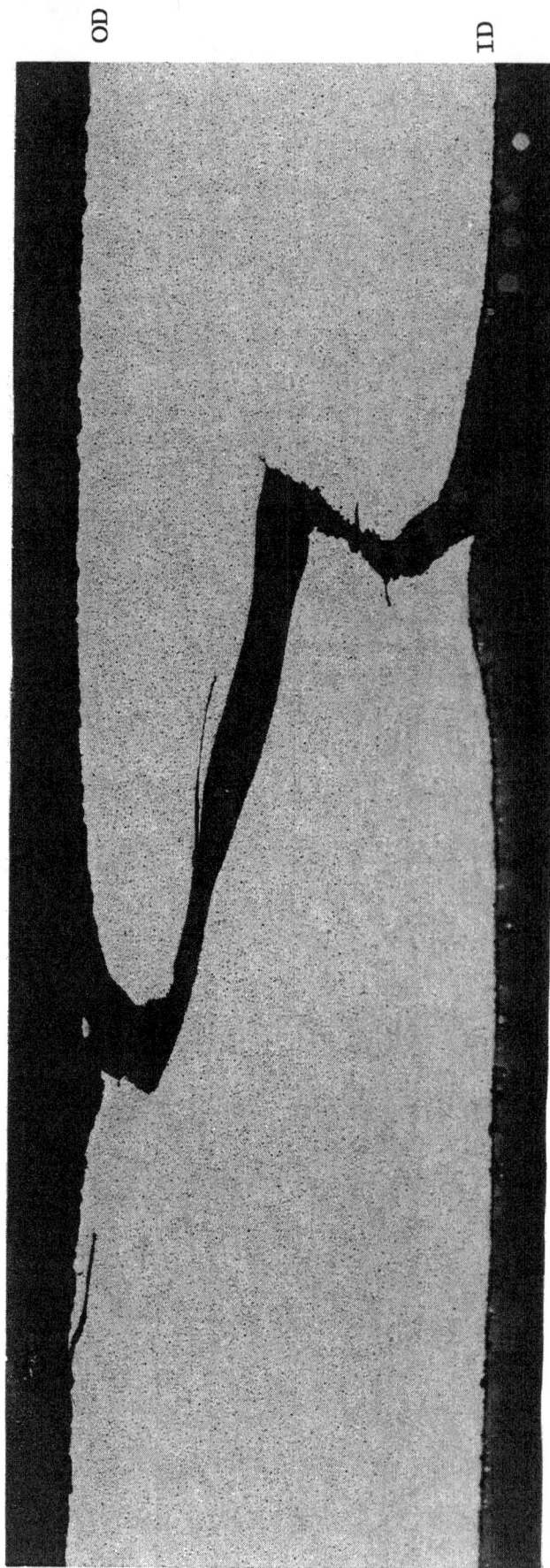

OD

ID

9E382

10X

Figure 3-14. A Seam Which Caused a Hydrostatic Test Failure.

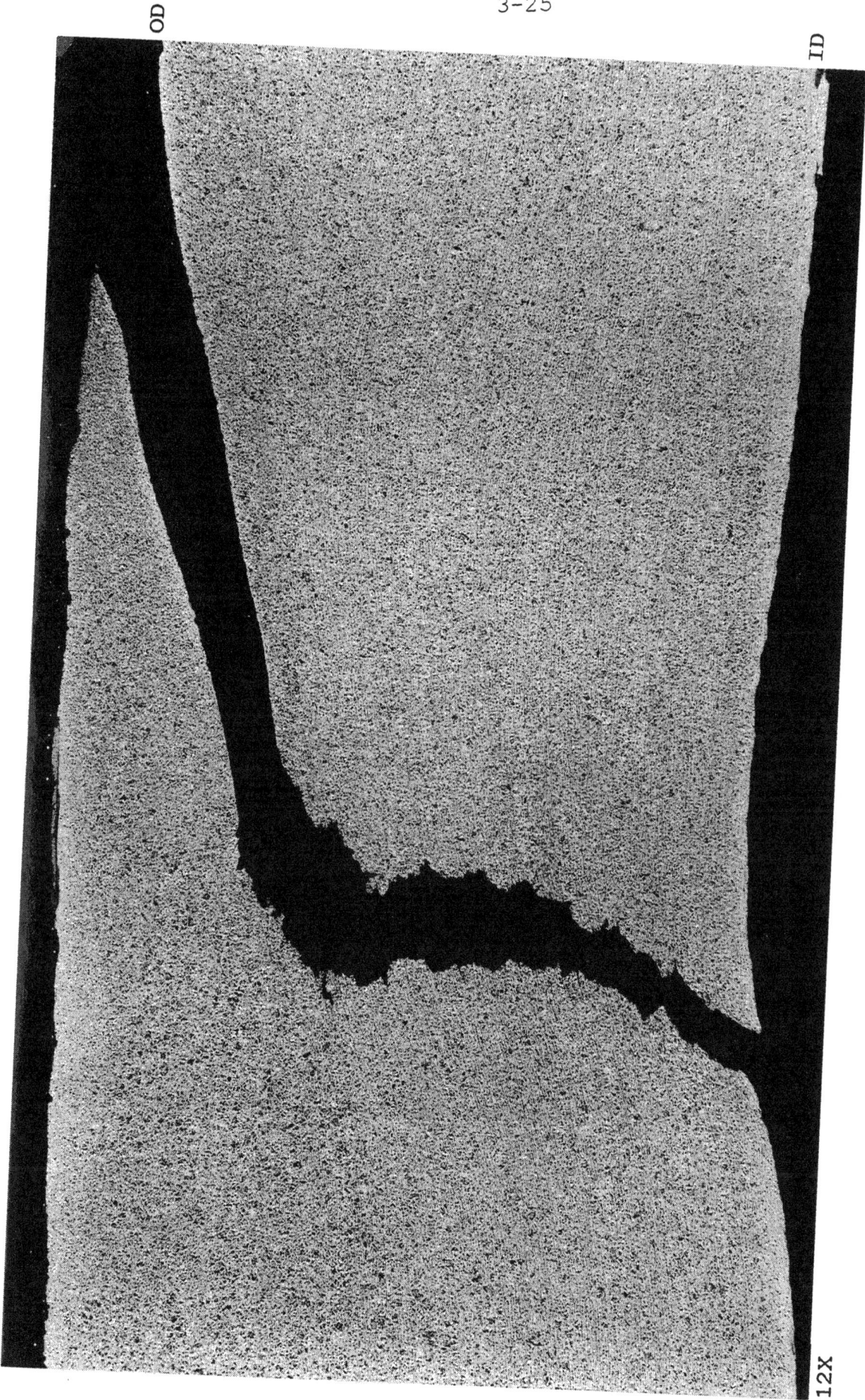

Figure 3-15. A Lap Which Caused a Hydrostatic Test Failure.

1X                                                    3G338

Figure 3-16.  Scabs and Blisters at the ID Surface.

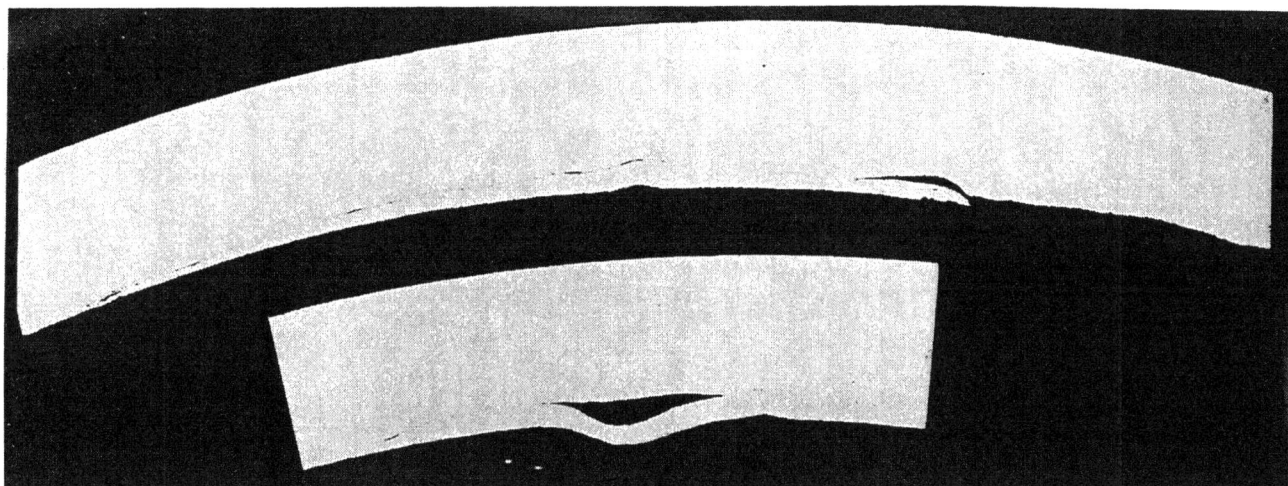

3X

3G340

Figure 3-17. Cross Sections of Scabs and Blisters.

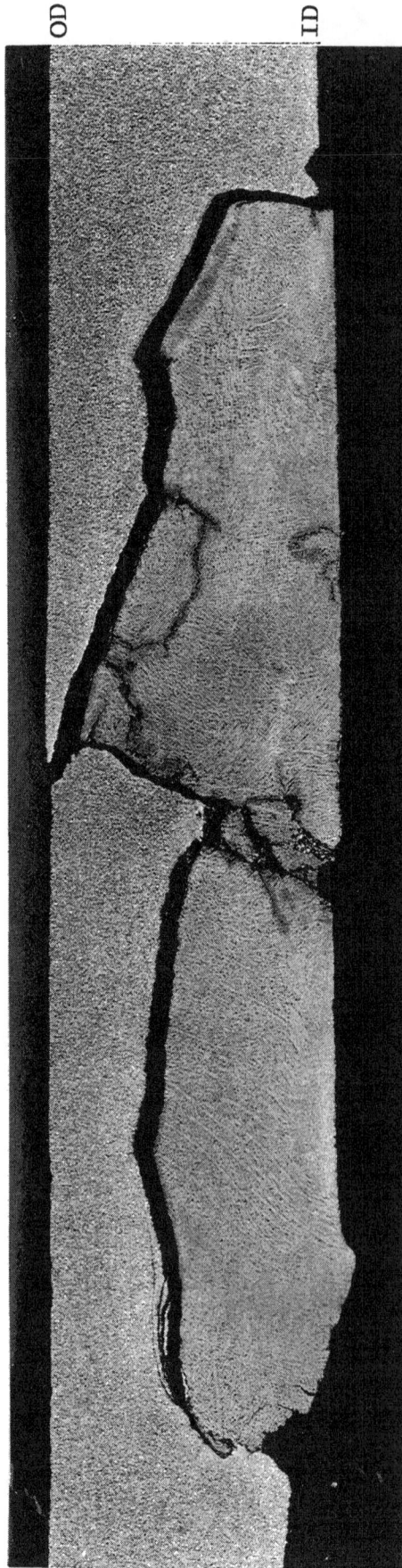

Figure 3-18. ID Roll-in.

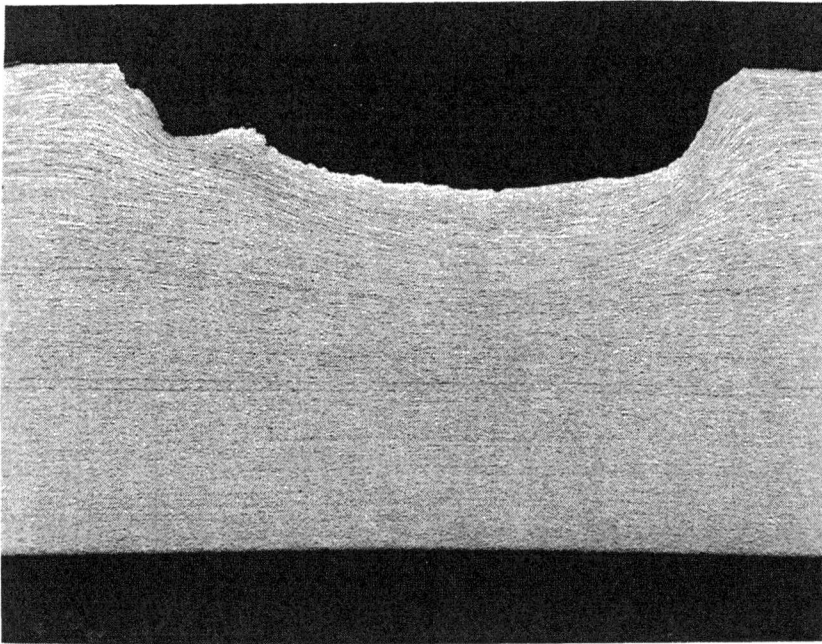

9X                                    IN927

Figure 3-19.   ID Pit.

# 4.0 ELECTRIC RESISTANCE WELDED PIPE

Electric resistance welded (ERW) line pipe is widely used for oil and gas pipelines. It is manufactured by cold-forming previously-hot-rolled strip to a circular shape, heating the two abutting edges by passing electric current through the interface as the edges come together, and effecting a bond between the edges as the molten or near-molten edges are forced together by mechanical means without the addition of any filler metal. The earliest ERW process (the "Parpart" process[4-1]) was developed in the 1910-1920 decade but the Johnson process which accounted for most of the ERW line pipe manufactured between 1930 and 1960 was invented in 1924[4-2]. The Johnson process, which involved low frequency a.c. current, 60 to 360 cycles per second, was almost entirely superseded by high-frequency ERW processes by 1970. A variation of the Johnson process involving d.c. current was used by at least one manufacturer between 1930 and 1970.

Described below is the general process by which ERW line pipe is made. The variations that existed in the past and those that are found in practice today are also explained, and the issues which affect ERW line pipe quality are discussed.

## The Process

ERW pipe is most commonly made continuously from hot-rolled strip steel which has been coiled and allowed to cool to ambient temperature although it has also been made in more than one instance[4-3,4-4] from single hot-rolled plates by cold forming the plates. Since hot-rolled strip is not made in all of the possible widths to accommodate the wide range of line pipe diameters, coils are often uncoiled, slit into two or more widths as required for a given pipe diameter and recoiled before being set up in an ERW production line.

Most ERW pipe production lines begin with uncoiling and leveling (flattening) the skelp. To provide for uninterrupted production a traveling loop of skelp is usually used so that the stationary end of one coil can be welded to the start of the next-coil while skelp continues to move uniformly toward the forming and welding stands. A typical arrangement is shown schematically in Figure 4-1. Not all mills utilize a travelling loop. In the absence of a "looper" the mill operation must be stopped to join the end of the next coil to the strip in the leveler. This can result in undesirable variations in quality.

Before the forming of the skelp begins, the edges are trimmed usually by rotary shears as shown in Figure 4-2. This creates the exact width of skelp needed to form the target diameter. In the older, d.c. and low-frequency a.c. mills the edges were also grit blasted following trimming to assure removal of all mill scale[4-5]. Otherwise contact resistance between the skelp and the rotary electrodes would have tended to prevent the passage of adequate welding current. With the advent of radio-frequency welding, however, this step became unnecessary. Rotary edge trimming in some mills is not the final trimming step. Some mills skive the edges with machine tools as a final step in edge preparation.

The flat, edge-trimmed skelp is then cold-formed into a tube by a series of rolls and/or other hardware[4-6,4-7]. In many cases the first series of rolls (concave-convex rolls) is similar to that shown in Figure 4-3. The skelp comes out of these rolls in a U-shape. The skelp is further deformed to a tube via either a series of adjustable internal and external cage rolls (Figure 4-4) or a series of vertical flanged rolls. The last stage of forming invariably involves "fin-pass" rolls as shown in Figure 4-5. The fin pass rolls also prevent the edges from contacting one another prematurely, and they may perform additional functions such as rounding the corners of the edges to prevent premature arcing and firmly positioning the edges to provide accurate alignment and to prevent buckling or waviness.

In a modern high-frequency conduction mill, the welding current is introduced through small sliding contacts and the physical bonding is effected by "squeeze" rolls and top pressure rolls as shown in Figure 4-6. In the older d.c. and low-frequency a.c. mills, the electrodes were large copper wheels which also supplied forming guides as shown in Figure 4-7. Another currently used welding technique consists of high frequency induction welding where the current is induced from an external coil concentric with the tube[4-8] (see Figure 4-8).

The production rates of older ERW mills ranged from 25 to 110 feet per minute but newer mills can produce pipe at speeds of up to 200 feet per minute. The driving force for moving the skelp is primarily supplied by two "pull-out" stands which pull the finished tube through the forming stands and the welder. However, some motive force is supplied to the uncoiled flattened skelp initially by "pinch" rolls in the vicinity of the uncoiler.

The appearance of an ERW seam immediately after being welded is as shown in Figure 4-9. The outside diameter and inside diameter flashes are trimmed by stationary tools before the weld cools. As the welded tube continues to move out of the welder and after the flashes have been trimmed, it passes through a seam annealer. The weld area is reheated by high-frequency induction coils to a temperature level intended to "normalize" the grain structure of the weld. This step typically results in grain refinement and it tends to enhance the toughness of the weld area. A trimmed and normalized weld is shown in Figure 4-10. Note the hour-glass shape of the weld heat affected zone, the distinct bond line and the upset pearlite-ferite layers. The wider darker area represents the effect of post weld heat treatment (normalizing).

After seam annealing the weld area is allowed to air cool to a temperature usually below 700°F as the tube moves toward a series of sizing and straightening rolls. Prior to entering this set of rolls the weld area is quenched with water. Depending on the particular manufacturer the seam may be given a mill-control ultrasonic or electro-magnetic inspection after being sized and straightened. At this point in most mills a rotary disc cut off or lathe-type cut off wheel moves along with the tube and cuts individual lengths of pipe. The individual lengths are marked with key information for tracking purposes and moved on for further processing.

Further processing consists of removing samples for required tests to meet the API 5L Specification, end preparation (usually beveling for welding), cleaning by high pressure water jets, hydrostatic testing, final visual and nondestructive inspection, weighing, measuring, and marking as required by the specification. The finished pipes may then be loaded directly onto barges, trucks or rail cars or diverted for other processing such as applying external and/or internal coating.

## Quality Factors

ERW pipe occasionally is unfairly stereotyped as being inherently inferior to either seamless pipe or DSAW pipe. In fact, when properly made it is as good as pipe made by any other process, and when certain special properties are required such as ultra-low carbon content and very high fracture resistance, it can be better than seamless pipe. Obtaining high-quality ERW pipe requires careful attention to skelp quality and to the pipe-making process variables. The sometimes-unfavorable reputation attached to ERW pipe arises from experiences involving either poor quality skelp or poorly controlled pipe manufacturing or both.  In particular, some ERW materials made by low-frequency welding processes and any ERW material made with "dirty" steel (containing high percentages of nonmetallic inclusions) can be plagued with defects which may lead to numerous hydrostatic test failures and occasional service failures.  Some ERW quality issues are discussed below.  In the case of issues that may affect pipe that has already been made or may be installed in an in-service pipeline, the intent is to help the purchaser or owner identify potential defects and to deal with them if possible.  In the case of issues that may affect pipe that is currently being manufactured or that is being sought, the intent is to help the purchaser identify ways to achieve a desired level of quality.

**Imperfections and Defects Associated with ERW Seams**

The ERW pipe imperfections and defects which may affect existing pipe or pipelines include
- Cold welds
- Penetrators
- Contact marks
- Lack of proper seam annealing
- Excessive trim
- Hook cracks
- Inadequate flash trim
- Inclusions
- Offset skelp edges
- Pinholes
- Stitching
- Weld area cracks
- Transit fatigue (see discussion of transit fatigue on Page 6-5).

## Cold Welds

A cold weld is defined by API Bulletin 5T1[4-9] as "A metallurgically inexact term generally indicating a lack of adequate weld bonding strength of the abutting edges, due to insufficient heat and/or pressure. A cold weld may or may not have separation in the weld line". A cold weld may result from too little heat or too little welding pressure or may also result from excessive welding pressure that causes so much extrusion that the metal at the interface is too cold to bond. Cold welds were more common in older d.c. and low-frequency welded materials than they are in modern high-frequency-welded materials because the welding current was more easily interrupted in those materials by contact resistance than it is for radio-frequency welded materials. A metallographic section and two fracture surfaces involving bondline ruptures which reveal the characteristics of cold welds are shown in Figure 4-11. The lack of a heat affected microstructure in Figure 4-11(a) suggests that the welding current (i.e., heat) was too low to effect bonding. The fracture surfaces shown in Figure 4-11(b) (not the same weld as Figure 4-11(a)) reflect intermittent cold welds (dark elliptically-shaped areas) interspersed with "stitching" (see definition below). Bondlines with intermittent cold welds and stitching of the type illustrated in Figure 4-11(b) may be surprisingly strong. It took a hydrostatic test to 107 percent of SMYS to get this one to fail. However, the initiating defects are small and the fracture is quite brittle illustrating the relatively low fracture toughness that accompanies such conditions.

## Cold Welds and Penetrators in High-Frequency Welded ERW Materials

With the advent of radio-frequency (400 kHz) seam welding, the contact resistance problems which often led to cold welds in low frequency or d.c. welded seams tended to disappear. However, the complex physics involved in high-frequency welding requires careful control of the process. Otherwise bondline defects peculiar to high-frequency phenomena may develop. The phenomena which lead to cold welds and "penetrators" in high-frequency-welded ERW seams were studied and documented by means of high speed cameras in References 4-10 and 4-11.

Proper bonding in a high-frequency ERW process is thought to result when welding takes place at the convergerence of the "vee" where the edges of the skelp meet. The authors of References 4-10 and 4-11 showed that cold welds arise when oxides are trapped in the converging vee preventing metal-to-metal bonding. This can occur even when welding is taking place at the vee-convergence as it is supposed to do. The oxides may not be properly excluded if the heat input is too low or there is too little mechanical upset. Penetrators, on the other hand, arise from welding not occurring at the vee-convergence. Instead the electromagnetic field may force molten metal out the convergence creating a narrow gap. The molten metal tends to bridge the gap at a "weld point" some distance behind the vee-convergence leaving an unwelded area. Penetrators tend to form when the speed of the skelp is too slow relative to the power input. At the threshold of penetrator formation, the actual nonwelded zones may be small and intermittent, but as the ratio of speed to power decreases the resulting nonbonded areas may grow in size. Also, in the latter case they tend to become periodic along the seam as described in Reference 4-10. Reference 4-12 describes the importance of the manganese to silicon ratio in the formation of penetrators. The association of cold welds and penetrators with predictable welding conditions indicates the essential need for controlling process variables in the making of ERW pipe in order to obtain a high-quality product.

## Contact Marks

Contact marks are defined by API bulletin 5T1 as "Intermittent marks adjacent to the weld line resulting from the electrical contact between the electrodes supplying the welding current and the pipe surface". The contact marks for the weld shown in Figure 4-10 are the very shallow dark-etched areas at the outside diameter surface on either side of the weld zone. Typically, these are quite superficial for high-frequency welds. As shown in Figure 4-12 the contact marks for a d.c. weld (and for low frequency welds as well) tend to be more pronounced in their effect on the microstructure. They correspond to the semi-circular dark-etched patterns at the outside diameter surface on either side of the bondline, in Figure 4-12. Also, they are accompanied on occasion by shallow surface melting.

## Lack of Proper Seam Annealing

If the seam is inadequately or incompletely reheated after welding, it may tend to exhibit excessive hardness or low toughness or both. This condition is sometimes caused by twisting or rotating of the pipe as it comes out of the welder. If it is allowed to rotate excessively the seam annealer may "miss" the weld line.

**Excessive Trim**

Excessive trim of the inside diameter flash is defined in the API Specification 5L. Trim which cuts below the inside diameter surface of the pipe as shown in Figure 4-13 is not necessarily excessive. In the case shown (0.500-inch wall thickness pipe), the trim is 0.02-inch. It would have to be 0.025-inch to be excessive.

**Hook Cracks**

A hook crack or upturned fiber imperfection is defined in API Bulletin 5TL as "Metal separations resulting from imperfections at the edge of the plate or skelp, parallel to the surface, which turn toward the I.D. or O.D. pipe surface when the edges are upset during welding". Hook cracks are not a welding problem per se although they do not exist other than at an upset weld such as an ERW seam. They arise from nonmetallic inclusions or laminations in the skelp that normally are parallel to the surfaces and do not affect the tensile strength of the skelp. The shear stresses between the layers as the fibers are bent cause the nonmetallic layers to rupture resulting in hook-shaped or J-shaped cracks near the bondline. Sometimes the cracks do not occur until the pipe is subjected to a large internal pressure such as in the mill or field hydrostatic test. The characteristics of hook cracks are illustrated in Figure 4-14. Note how the cracks shown in Figure 4-14(a) follow the upturned fiber pattern near the bondline. The layered or "woody" fracture surface shown in Figure E-14(b) is commonly associated with hook crack ruptures. Hook cracks not exposed by a hydrostatic test would seldom be expected to cause problems in service unless extended by fatigue crack growth from large numbers of significant-size pressure cycles.  Hook crack failures during the retesting of older ERW pipelines are fairly common.  Hook cracks can be prevented or minimized if a low-sulfur skelp (0.003 percent or less) is used to make ERW pipe.  A low sulfur content or the addition of alloys which tie up sulfur in hard spherical particles limits the formation of MnS inclusions which elongate upon rolling to form the brittle precursors of hook cracks.

**Inadequate Flash Trim**

The ID flash shown in Figure 4-12 was not trimmed flush but it would not be considered excessive even by today's API 5L requirements.  Inadequately trimmed flash tends to create edges which reflect ultrasonic signals and hence it can significantly interfere with nondestructive inspection of the seam.

**Inclusions**

Inclusions typically refer to the nonmetallic compounds found as impurities in steel.  These can be harmful in two respects.  They are precursors to hook cracks if they exist in large quantities at the edges of the skelp used to form ERW pipe.  Secondly, they tend to reduce the ductile fracture resistance of the material.

**Offset Skelp Edges**

An example of slightly offset skelp edges is shown in Figure 4-12.  The skelp-edge on the right side of the bond line was displaced radially outward from that on the left side.  Note that the bondline is deflected on an angle because of the offset edges. Because the API 5L specification allows up to 0.060-inch of offset (including trim), the amount of the offset of the edges of the pipe in this photograph is within the allowable limit. Significantly, worse offsets have been observed in older pipelines, and in at least one case they have resulted in the development of a fatigue crack and a failure. Offset edges may result from cambered skelp or wavy skelp edges. When it occurs frequently in an order of pipe, it suggests that something is wrong with either the coils of skelp or the tube forming equipment.

## Pinholes

A pinhole is defined in API Bulletin 5T1 as "A short unwelded area in the weld line extending through the entire pipe thickness so that fluid will leak out through the area very slowly".

## Stitching

Stitching is defined in API Bulletin 5T1 as "Variation in the properties of the weld occurring at short regular intervals along the weld line due to repetitive variation in welding heat. The variation in properties gives rise to a regular pattern of light and dark areas visible only when the weld is broken in the weld line". An example of stitching is visible on the fracture surfaces shown in Figure 4-11(b). It is not the elliptically-shaped dark areas, those are cold welds. It is the faint repeated vertical patterns which extend through the wall thickness in some areas and which appear on both the fractured and cold welded surfaces. Stitching is a phenomenon associated with low frequency ERW seams. It occurred when the welding speed was sufficiently fast that the 60 or 120 cycle power fluctuations were causing a repeated variation in welding heat. It would not be expected in either d.c. ERW or radio-frequency ERW pipe. Stitching is usually accompanied by poor toughness (i.e., brittleness) along the bond line.

## Weld Area Cracks

Cracks in the weld area could possibly arise for a number of reasons in that they are a "stress-induced separation of the metal". If the weld area is heated excessively, spring back stresses may cause cracking of the weld zone before it cools sufficiently. Cracks may occur in residual unsound metal if there is insufficient flash extrusion. In addition, cracking may occur later as the result of process malfunctions in the manufacture of ERW seams. For example, if the post-weld heat treatment is insufficient or if the weld becomes misaligned so that the heat treatment misses the weld, untempered martensite may form. This material has been observed to become cracked after the pipeline has been in service, ostensibly because of hydrogen charging from cathodic protection. A example of such cracking is shown in Figure 4-15.

### Selective Corrosion of ERW Seams

Some ERW seam materials are susceptible to selective or "grooving" corrosion as shown in Figures 4-16 and 4-17. This phenomenon can occur when the pipe is undergoing either external or internal corrosion-caused metal loss and the metal loss affects the area of the ERW seam. As suggested by Figures 4-16 and 4-17, the corrosive action can preferentially attack the bondline region at a higher rate than the surrounding material.

The result is often the formation of a v-shaped groove centered on the bondline. The facts that the resulting corrosion produces a relatively sharp notch as opposed to a blunt pit and that the bondline region in older ERW materials is not as tough as the parent material, indicate that this is a serious form of deterioration that is likely to cause a rupture much sooner than might be the case for corrosion in the parent material.

The phenomenon of selective seam corrosion is described in more detail in References 4-13 through 4-19.

## Material Factors and Quality Control

All major ERW line pipe producers in North American currently utilize high frequency welding processes. Thus, one can expect cold welds to be less of a problem than might have been the case in older low-frequency or d.c. welded ERW materials. Also, stitching would not be expected in the newer materials. However, to avoid some of the other problems such as hook cracks, offset skelp edges, or selective corrosion susceptibility, a potential purchaser must be attentive to certain material factors and processing steps. It is possible to eliminate most of the potential problems by specifying reasonable levels of Charpy V-notch impact transition temperatures and absorbed energy.[4-20] To obtain reasonable levels, the manufacturer will have to make or purchase high-quality skelp that will undoubtedly have low carbon and low sulfur contents (i.e., both less than 0.010 percent by weight). A more elusive but better target for sulfur would be a maximum of 0.003 percent. Clean, low-sulfur steels greatly reduce the chances that elongated nonmetallic inclusions will be formed. The latter are frequent precursors to hook cracks. Alternatively, sulfide shape control can be effected by specific alloying elements. These elements form compounds with sulfur which have softening temperatures above the hot rolling temperature range for steel. However, the benefits of low sulfur extend to increased resistance to selective corrosion.[4-10] Low carbon, of course, contributes to weldability.

In addition to skelp quality the pipe-making process details are important. Factors which can result in improved quality of the product are maintenance of the skelp edge trimming equipment; attention to the forming rolls relative to the diameter, wall thickness, and tensile properties of the skelp; whether or not the skelp is slit and where it is slit (i.e., the center of the coil is less desirable than the ⅓ points); the use of strand cast slabs to eliminate center macrosegragation, maintenance of the alignment of the squeeze rolls, maintenance of the welding equipment and post weld heat treatment equipment. It is possible to monitor the effects of these factors on quality by periodically extracting and examining a metallographic section across the seam.

Finally, the results of inspection and testing are vital to judge the quality of the product. Visual inspection and dimensional checks as required by the API Specification 5L should be monitored and spot-checked by the purchaser or his representative. The nondestructive seam inspection should also be monitored and the number and nature of any rejects should be ascertained. Some manufacturers will agree to a mill hydrostatic test to 100 percent of SMYS for as long as 20 seconds as long as plastic deformation does not interfere with end seal integrity. This level of testing is superior to the required 10 second test to 90 percent at SMYS and it probably represents the most rigorous test feasible for the manufacturer.

# References

4-1.    Thomas, P. D., "Pipe-Manufacture and Use" *Iron and Steel Engineer*, July 1957.

4-2.    Heald. S. T., "Radio-Frequency Resistance Welding of Carbon Steel Pipe and Tubing", *Mechanical Working of Steel 1*, AIME, 1963.

4-3.    Kane, G.E., and Mendez, E., "High Frequency Welding of LOD Pipe", *Mechanical Working and Steel Processing*, VII, AIME, 1969.

4-4.    Morehead, C. W., "Operation of New 16-Inch Electric-Weld Pipe Mill", *Iron and Steel Engineer*, December, 1950.

4-5.    Keska, K. R., "Recent Advances in Electric Resistance Welding of Steel Tubes-Both AC and DC", *Mechanical Working and Steel Processing V,* Proceedings of the N_nth Mechanical Working and Steel Processing Conference, AIME, December 1966.

4-6.    Patterson, Billy R., "The Evolution and Development of ERW Pipe", Pipe Symposium, Midyear Meeting of API Committee on Standardization, June 1989.

4-7.    Rode, Heinz, "Electric Weld Pipe Production Today", *Iron and Steel Engineer*, October 1967.

4-8.    Koch, F. O., and Peters, P. A., "Distinguishing Characteristics of High-Frequency Induction-Welded Pipe", *3R International*, March 1987.

4-9.    API Bulletin 5T1, "Bulletin on Imperfection Technology", Ninth Edition, May 31, 1988.

4-10.   Haga, H., Aoki, K., and Sato, T., "Welding Phenomena and Welding Mechanisms in High-Frequency Electric Resistance Welding Study on ERW", Report 1, Nippon Steel Corporation, Presented at 60th Annual Meeting of the American Welding Society, 1979, *Welding Res. Supplement*, p 208-212 July 1980.

4-11.    Haga, H., Aoki, K., and Sato, T., "The Mechanisms of
         Formation of Weld Defects in High-Frequency Electric
         Resistance Welding", *Welding Res. Supplement*, p 104-
         109, June 1981.

4-12.    Yokoyama, E., Ejima, A., Watanabe, S., yoshimoto, y.,
         and Mirano, Y., "Effects of Welding Conditions and
         Mn/Si Ratio on the Penetrator Defect Occurrence in ERW
         High Manganese Line Pipe", Kawasai Steel Corporation,
         September, 1979.

4-13.    Kato, C., Otogura, Y., Kado, S., and Hisanatsu, Y.,
         "Grooving Corrosion in Electric Resistance Welded Steel
         Pipe in Sea Water", *Corrosion Science*, **18**, pp 61-74,
         1978.

4-14.    Masamura, K., and Matsushima, I., "Grooving Corrosion
         of Electric Resistance Welded Steel Pipe in Water Case
         Histories and Effects of Alloying Elements",
         *CORROSION/81*, Toronto, Ontario, Canada, April 6-10,
         1981.

4-15.    Heitmann, W. E., Southwick, P. D., and Pausic, F., "ERW
         Line Pipe: The Effect of Welding and Annealing Upon the
         Properties, Microstructure, and Corrosion Resistance",
         *HSLA Steels, Technology & Applications*, ASM, pp 957-
         966.

4-16.    "Corrosion of ERW Casing", Nippon Steel Corporation
         (C-85-4-1-E) November 1985.

4-17.    Duran, C., Treiss, E., and Herbsleb, G., "The
         Resistance of High Frequency Inductive Welding Pipe to
         Grooving Corrosion in Salt Water", *Salt Water
         Corrosion*, p 41-48, September 1986.

4-18.    Miyasaka, A., Ogawa, H., and Katoh, K., "Grooving
         Corrosin Behavior of HF-ERW Pipes and Tubes as
         Influenced by Petroleum production Environments",
         *CORROSION/91*, Cincinnati, Ohio, March 1-15, 1991.

4-19.    Katoh, K., Tanioka, S., Kosuge, N., hiroshi, Kihira,
         Ioth, S., Murata, T., and Miyasaka, A., "Evaluation of
         Grooving corrosion-Resistanct High Frequency Electric
         Resistance Welded (HF-ERW) Steel Pipes by New
         Electrochemical Technique", Nippon Steel Technical
         Report No. 39, 40 pp January 1989 (UDC620.193:541.13).

4-20.     Kiefner, J. F., and Maxey, W. A., "Specifying Fracture Toughness Ranks High in Line Pipe Selection", *Oil and Gas Journal*, October 9, 1995.

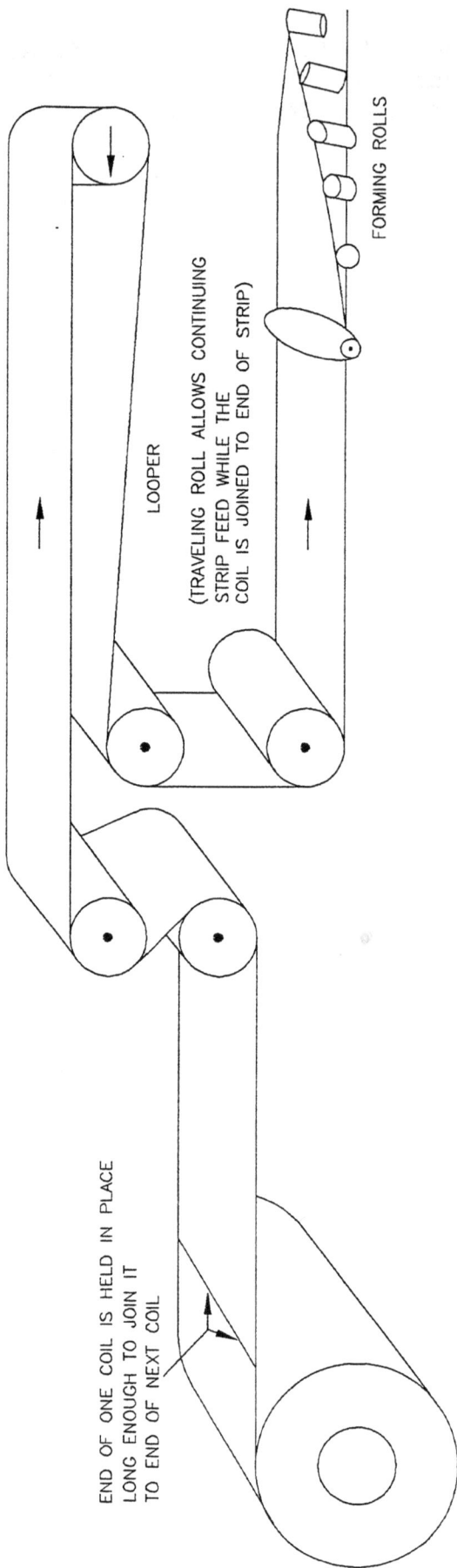

Figure 4-1. Looper Which Permits Welding of One Coil to the Next

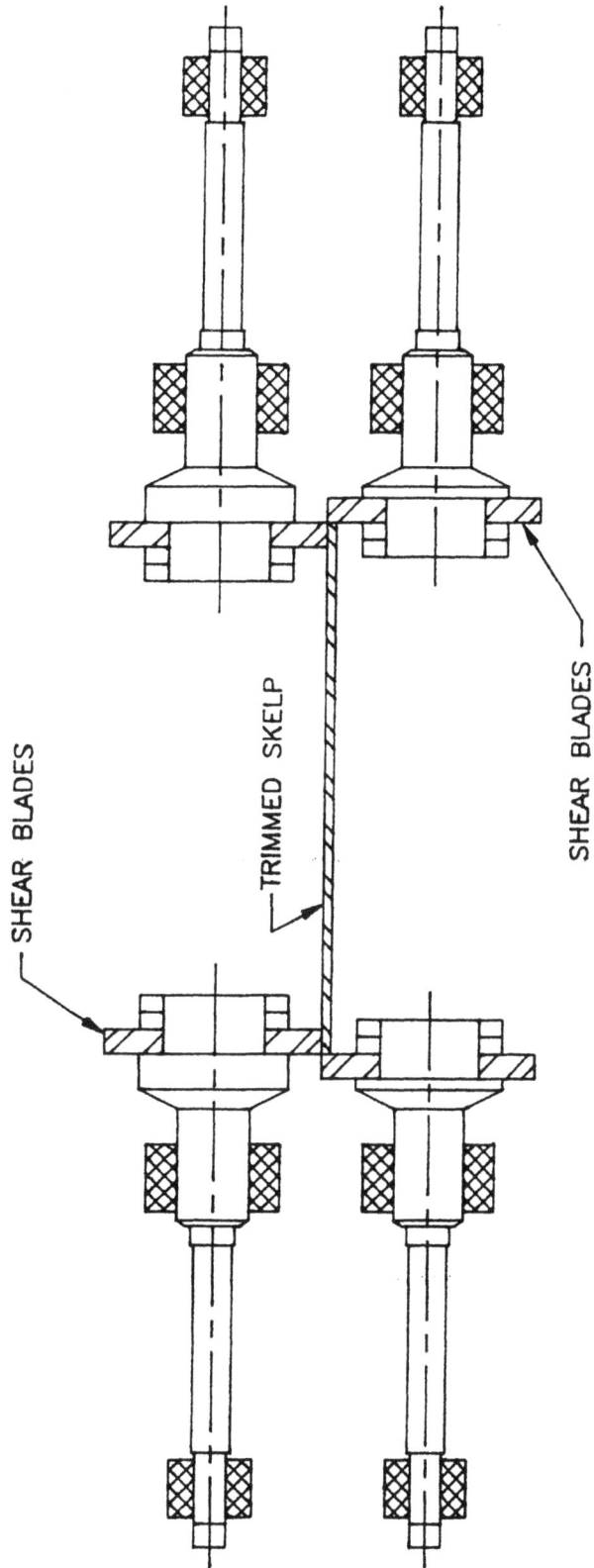

SHEAR BLADES

TRIMMED SKELP

SHEAR BLADES

Figure 4-2. Edge Trimming Rotary Shears

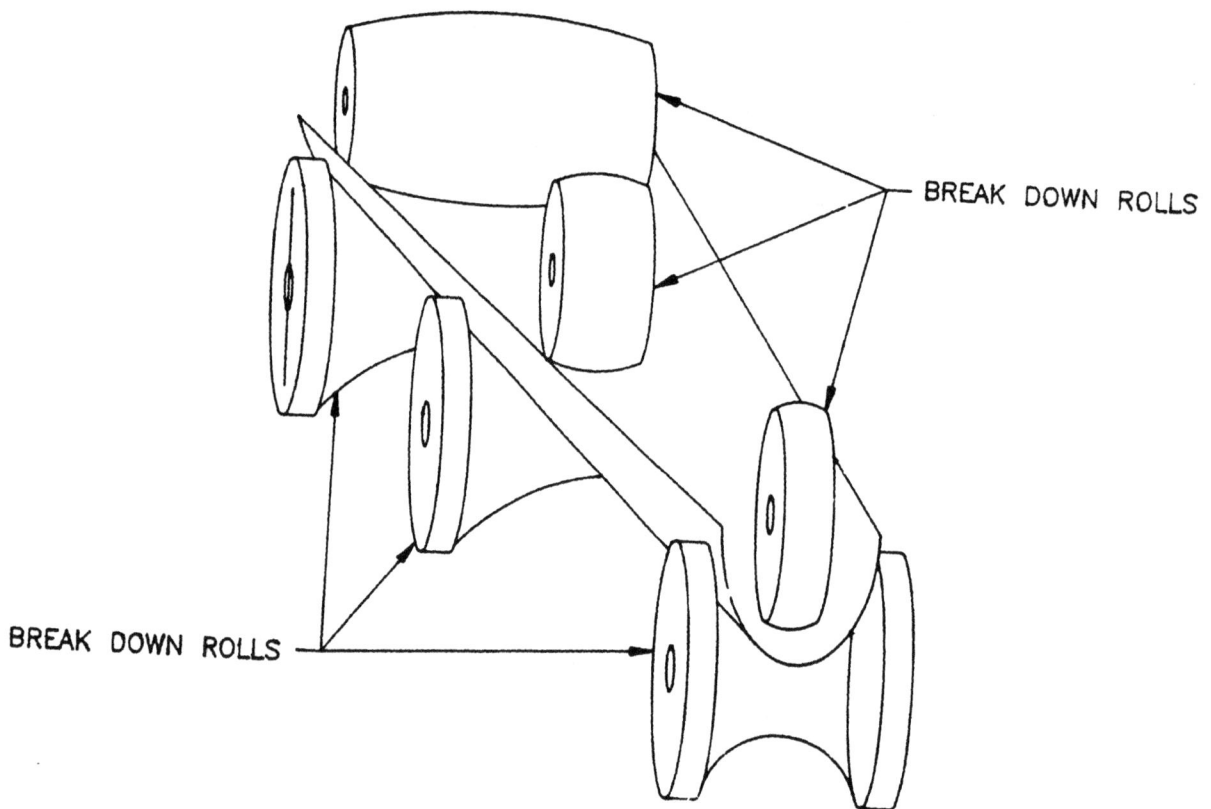

Figure 4-3.  Concave — Convex Roll Forming

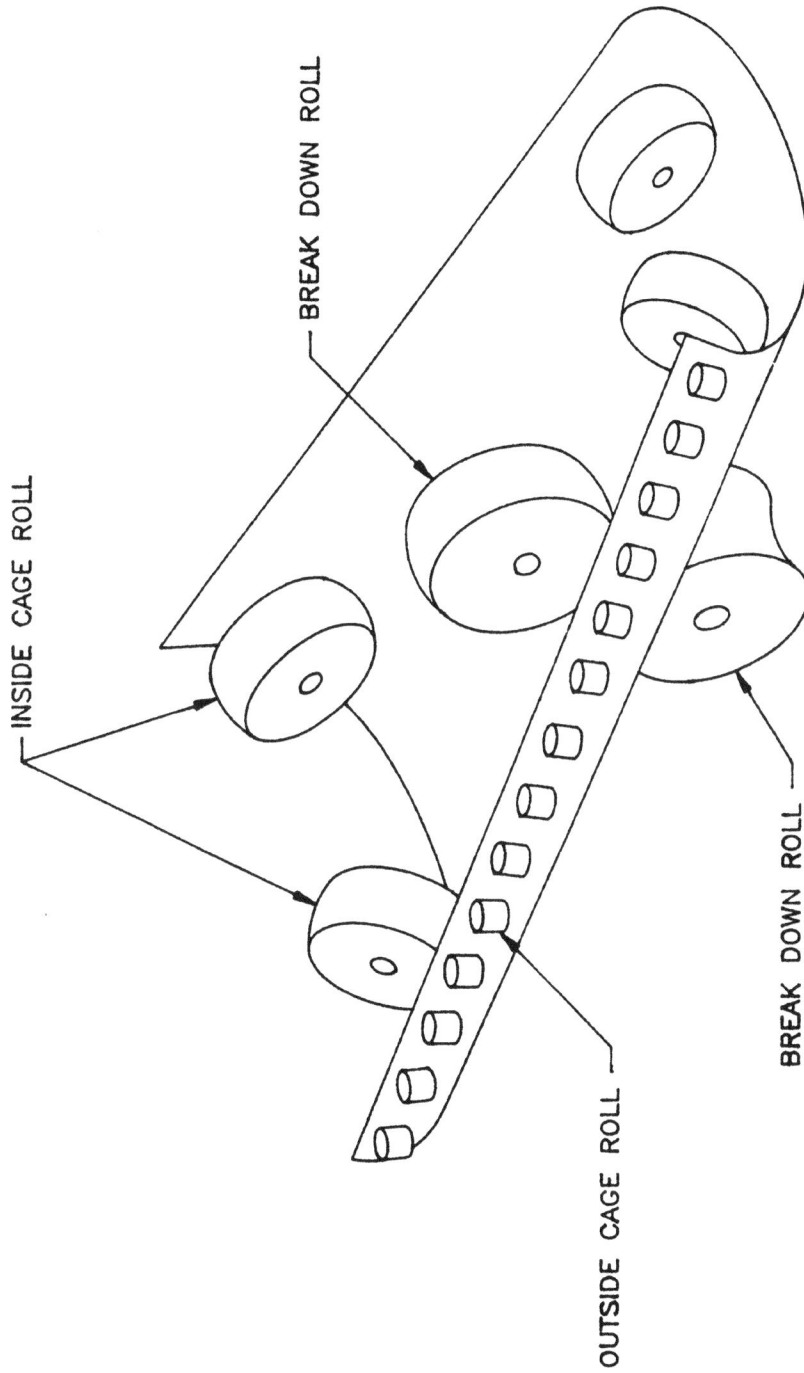

Figure 4-4. Cage Roll Forming

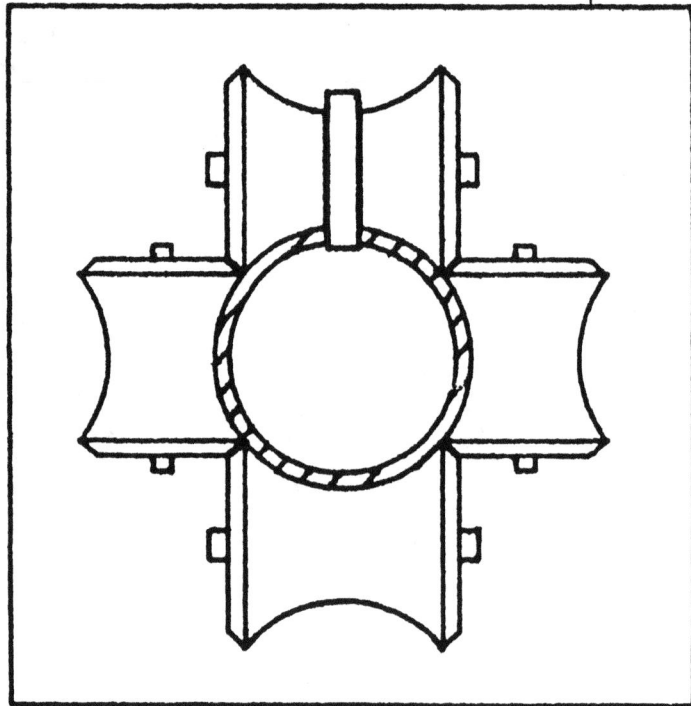

Figure 4-5.   Fin Pass

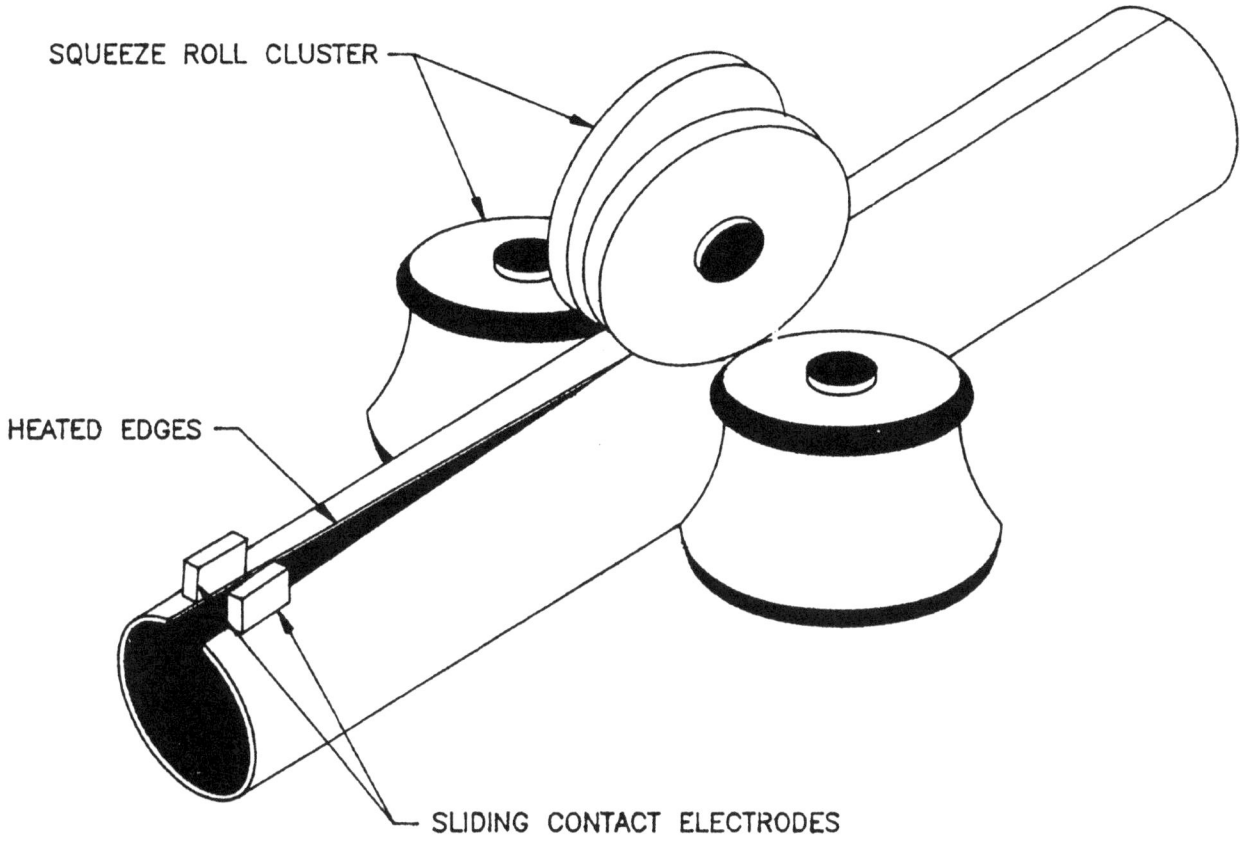

Figure 4-6.  High Frequency Welder

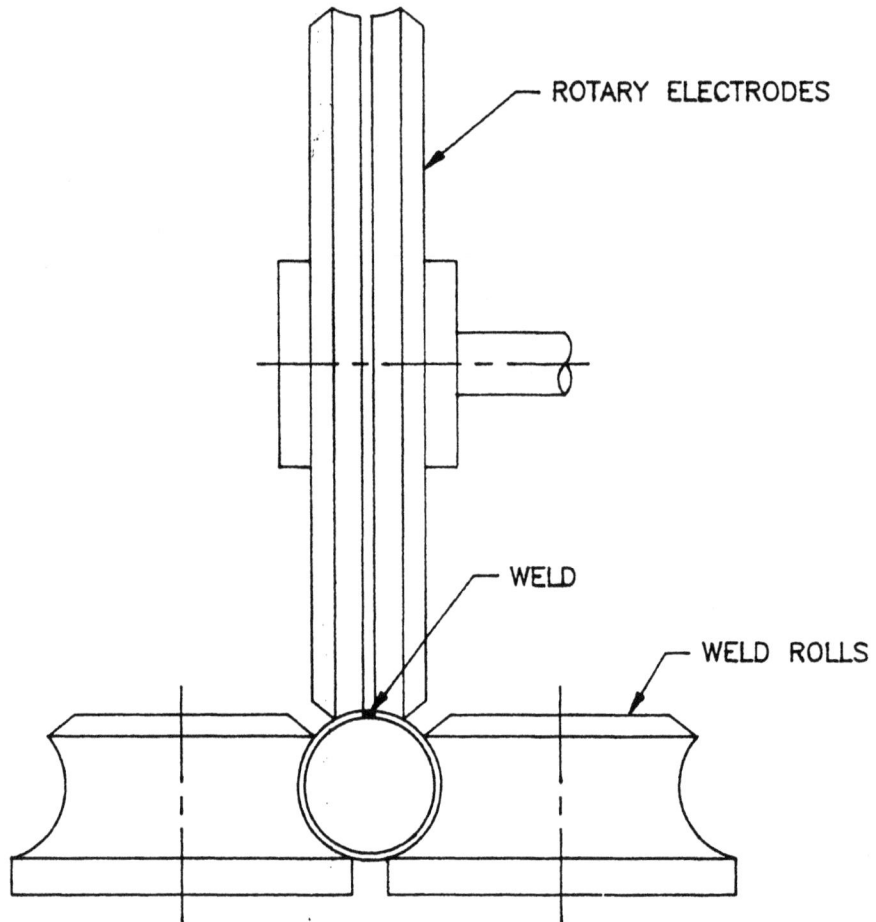

Figure 4-7.  Low Frequency Welder

WELD ROLL

COIL (3 TURN)

OPTIMUM COIL POSITION

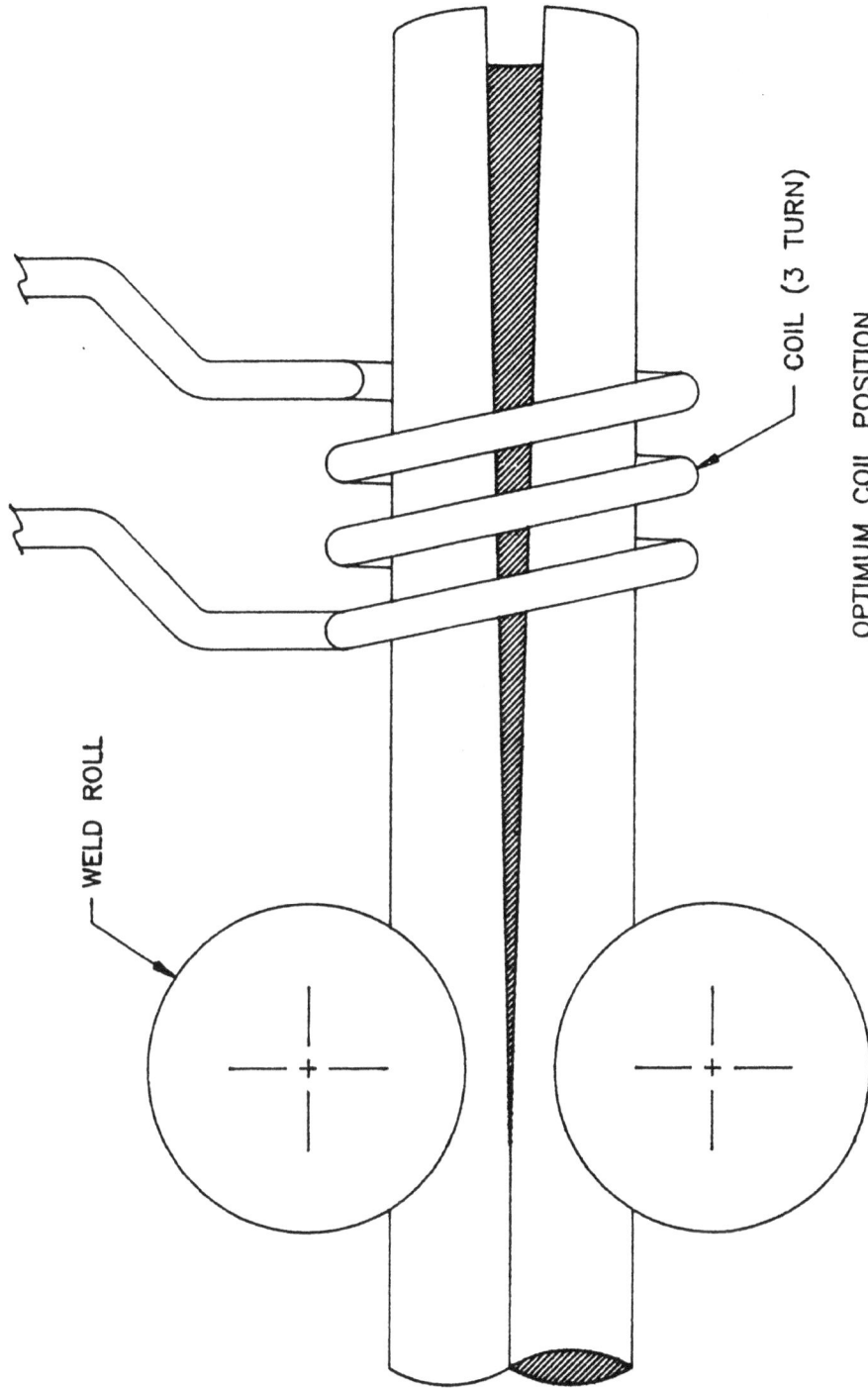

Typical three-turn coil design and position diagram for induction welding.

Figure 4-8.    Induction Welder

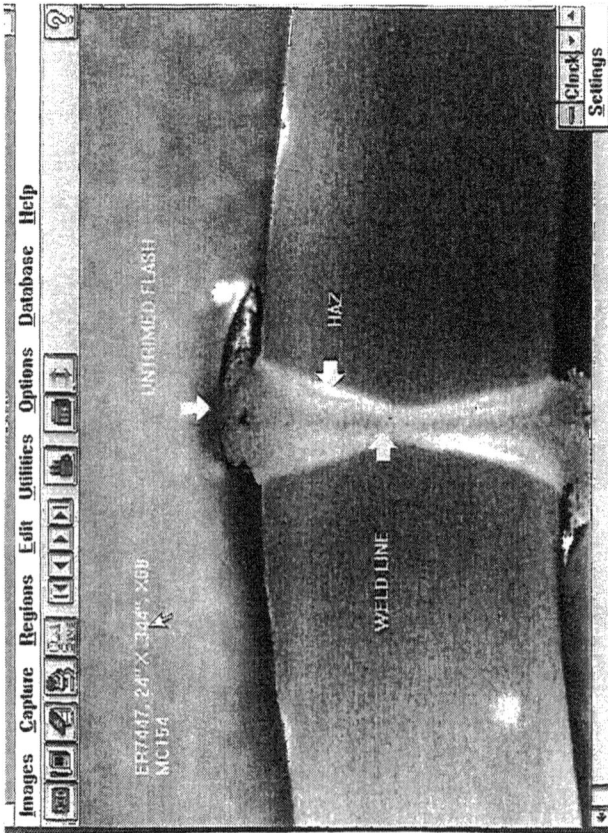

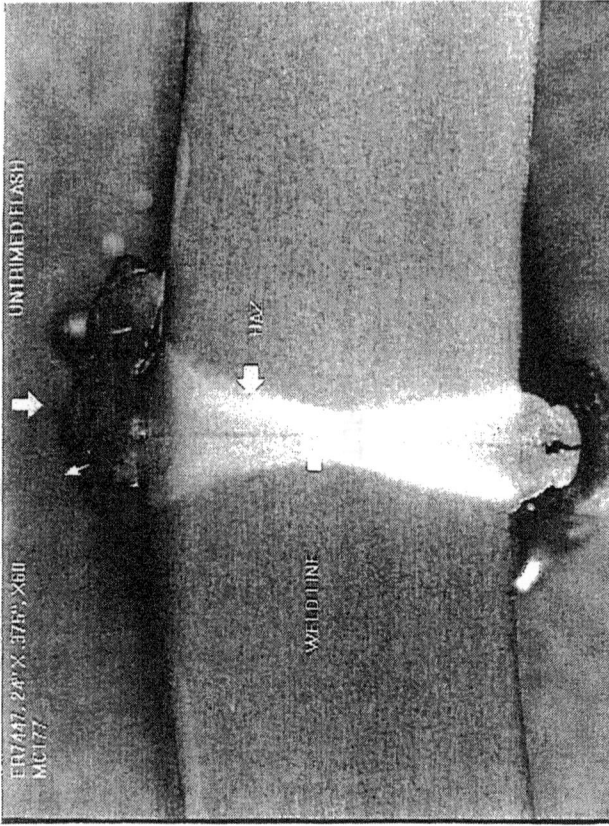

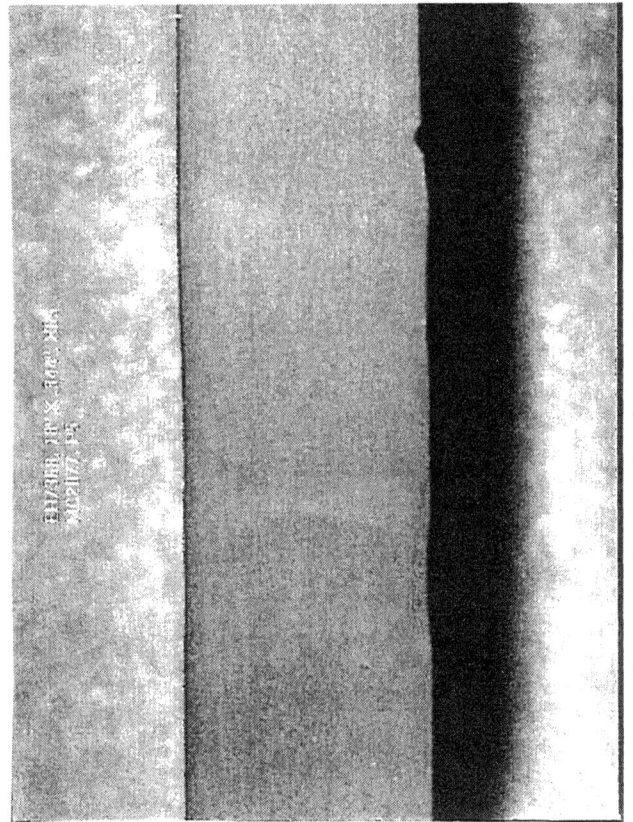

Figure 4-9.  Appearance of ERW Seam Before (top) and after (bottom) Trimming and Normalizing

OD

ID

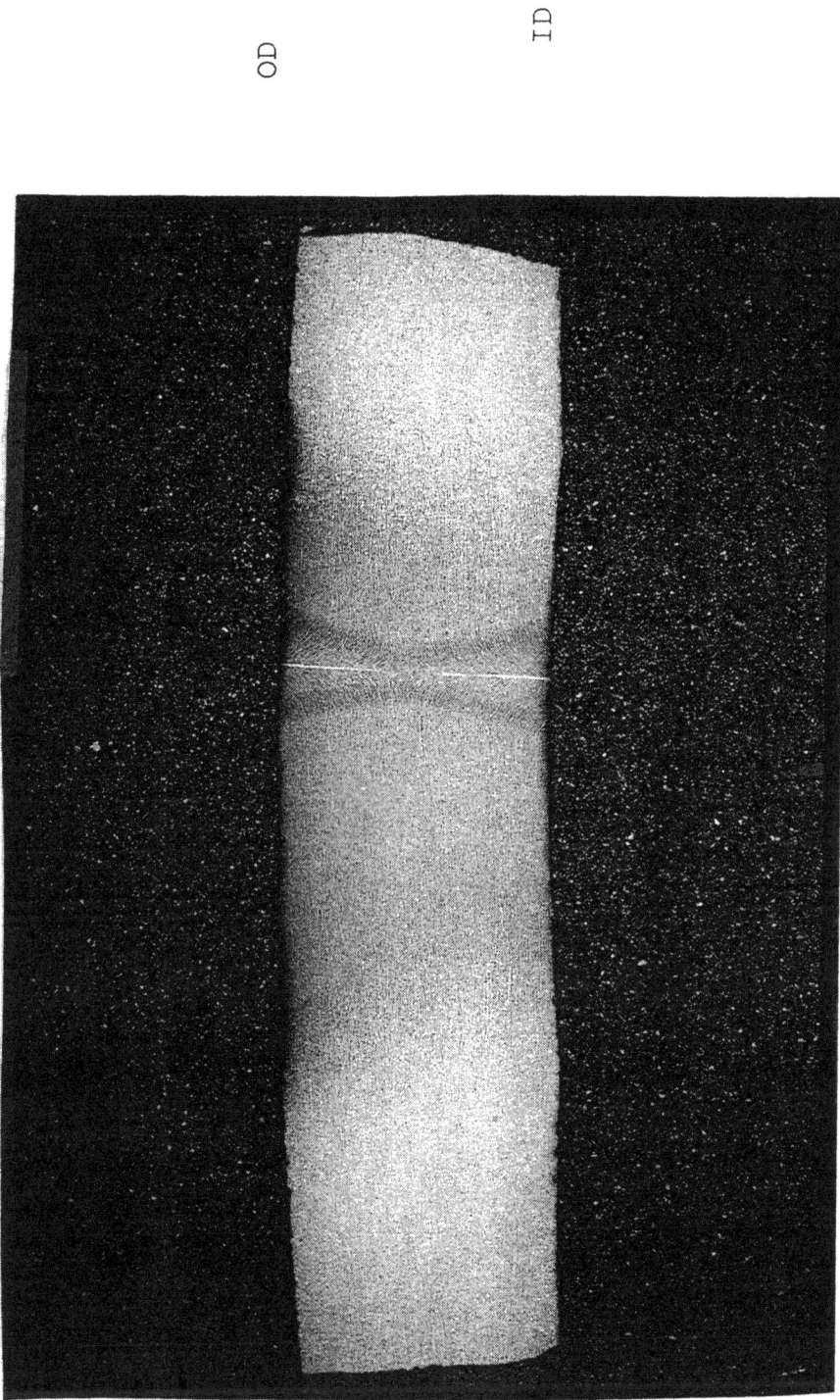

Figure 4-10. Metallographic Section Across Trimmed and Normalized High-Frequency ERW Seam

OD

ID

10X

(a)

(b) Outside surfaces back-to-back

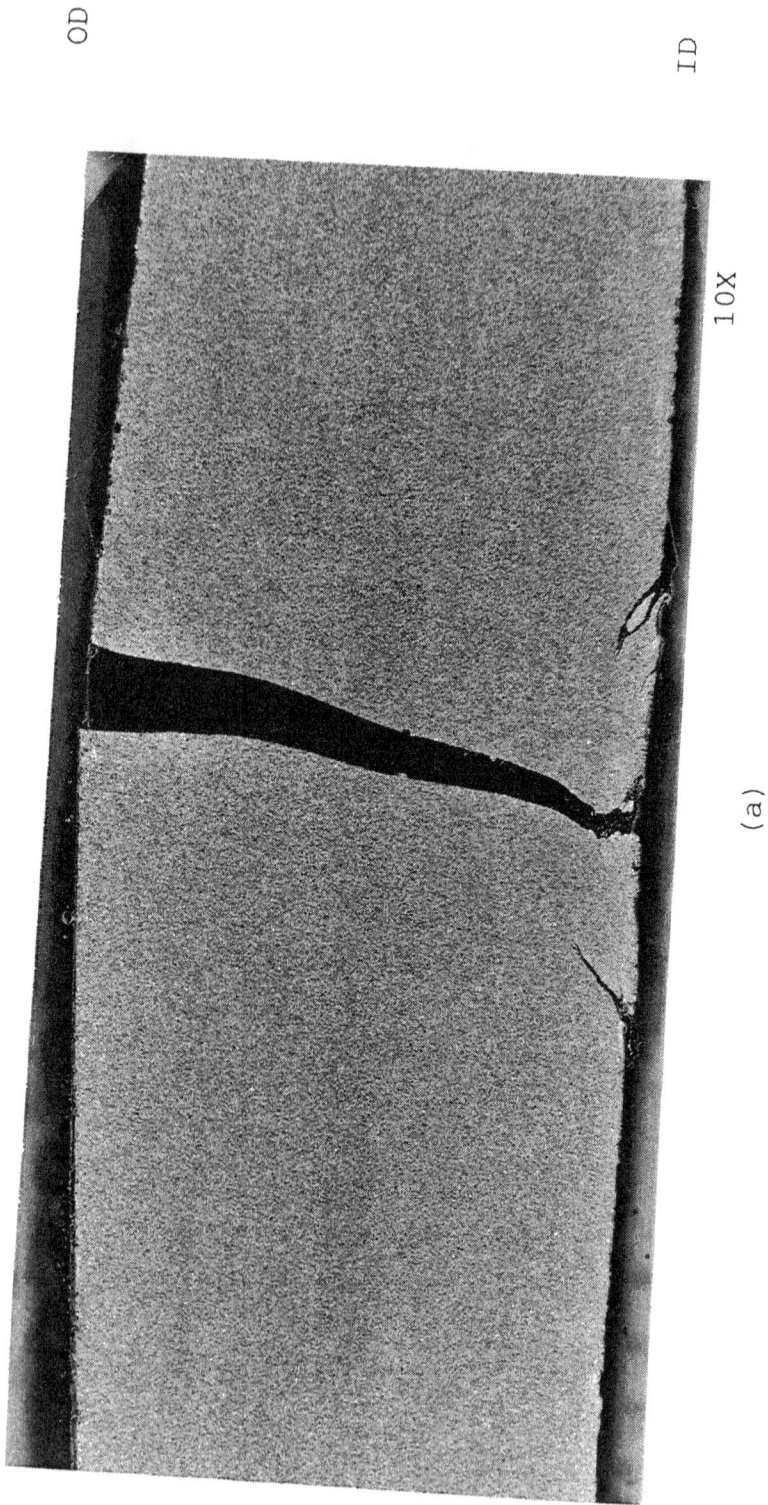

Figure 4-11. Characteristics of Cold Welds and Stitching

8X

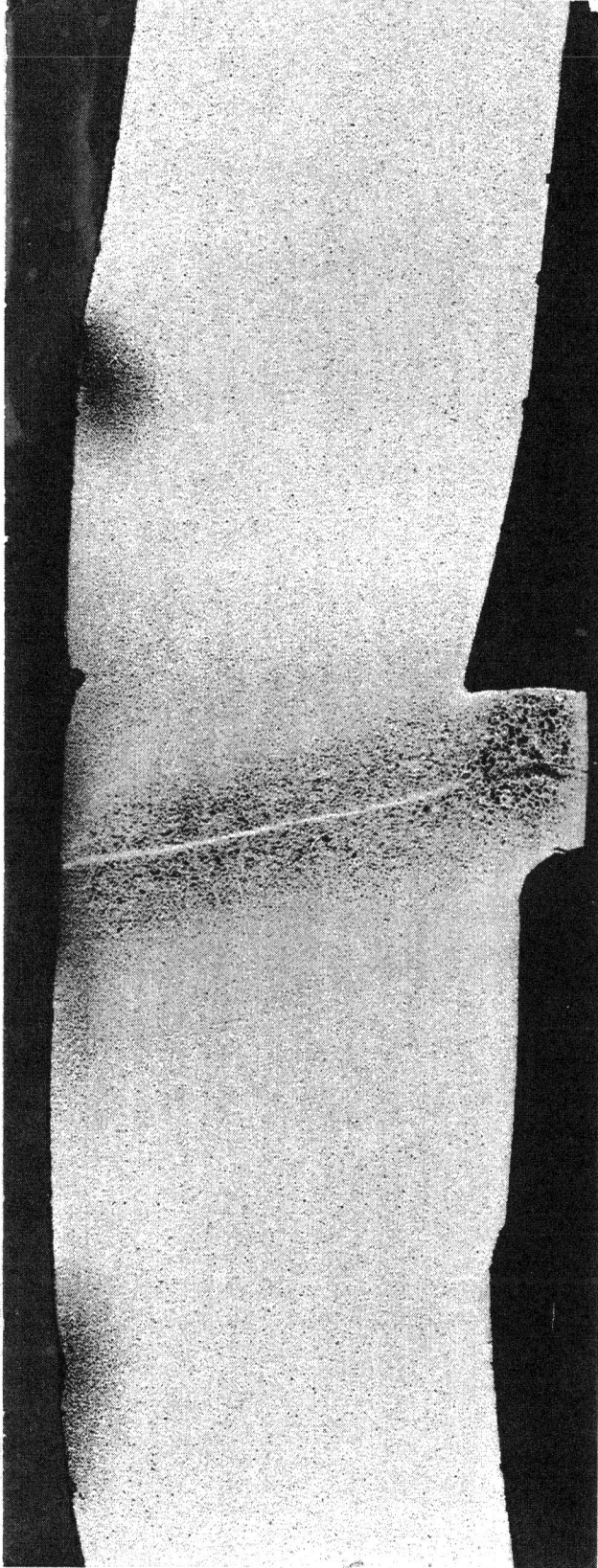

Figure 4-12.  A d.c.-Welded ERW Seam Showing Contact Marks and Offset Skelp Edges

OD

ID

5X

Figure 4-13.   ID Flash Trim Which Cuts Below Original
Pipe Surface But is Not Excessive

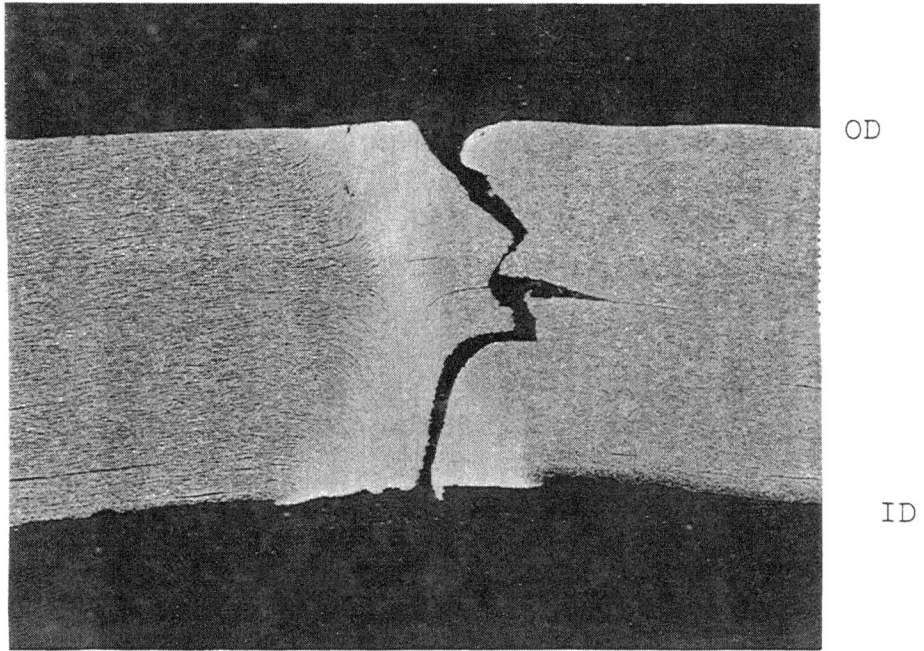

OD

ID

10X

(a)

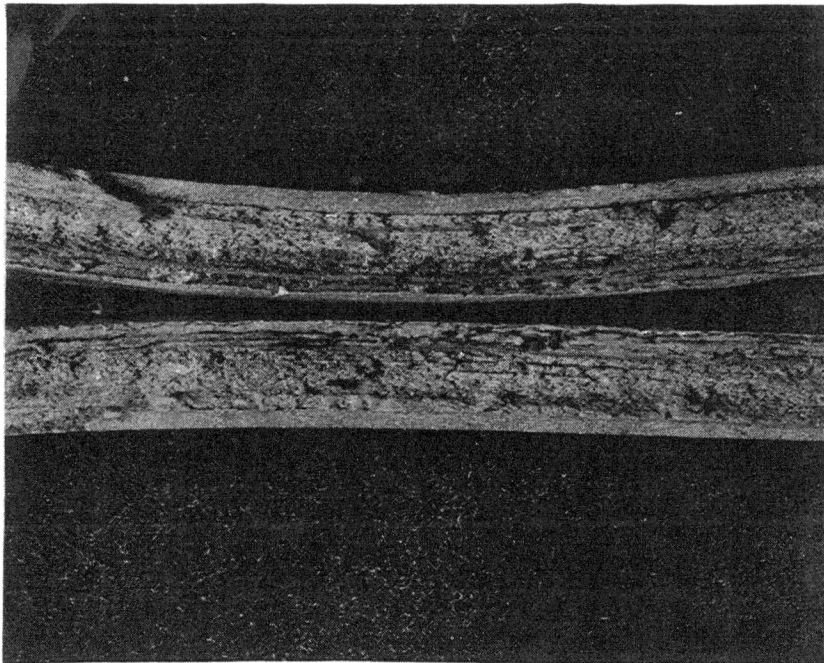

(b) Outside Surfaces back-to-back

Figure 4-14.  Characteristics of Hook Cracks; (a) Metallographic
Features, (b) Fracture Surface Appearance

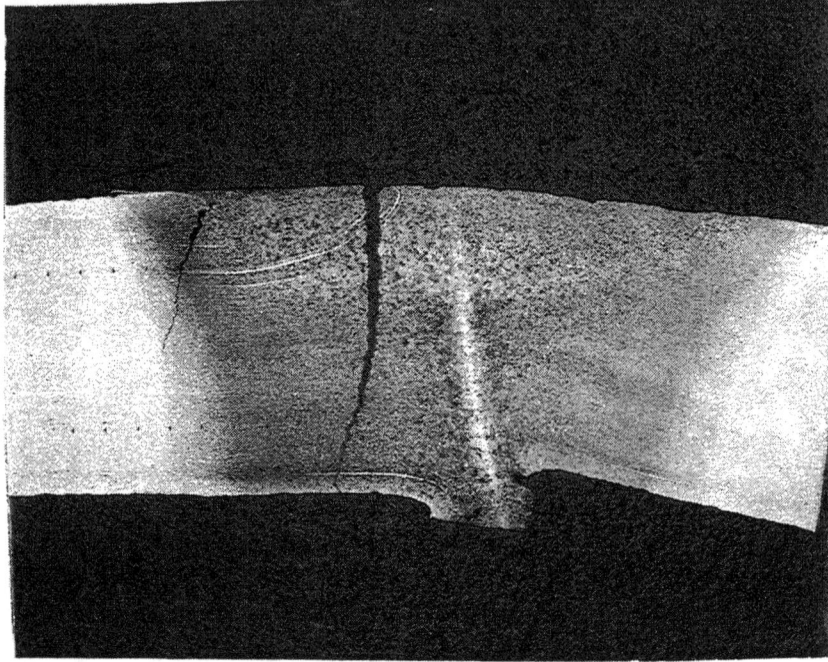

Figure 4-15.   Hydrogen Cracks of Excessively Hard Heat Affected
Material

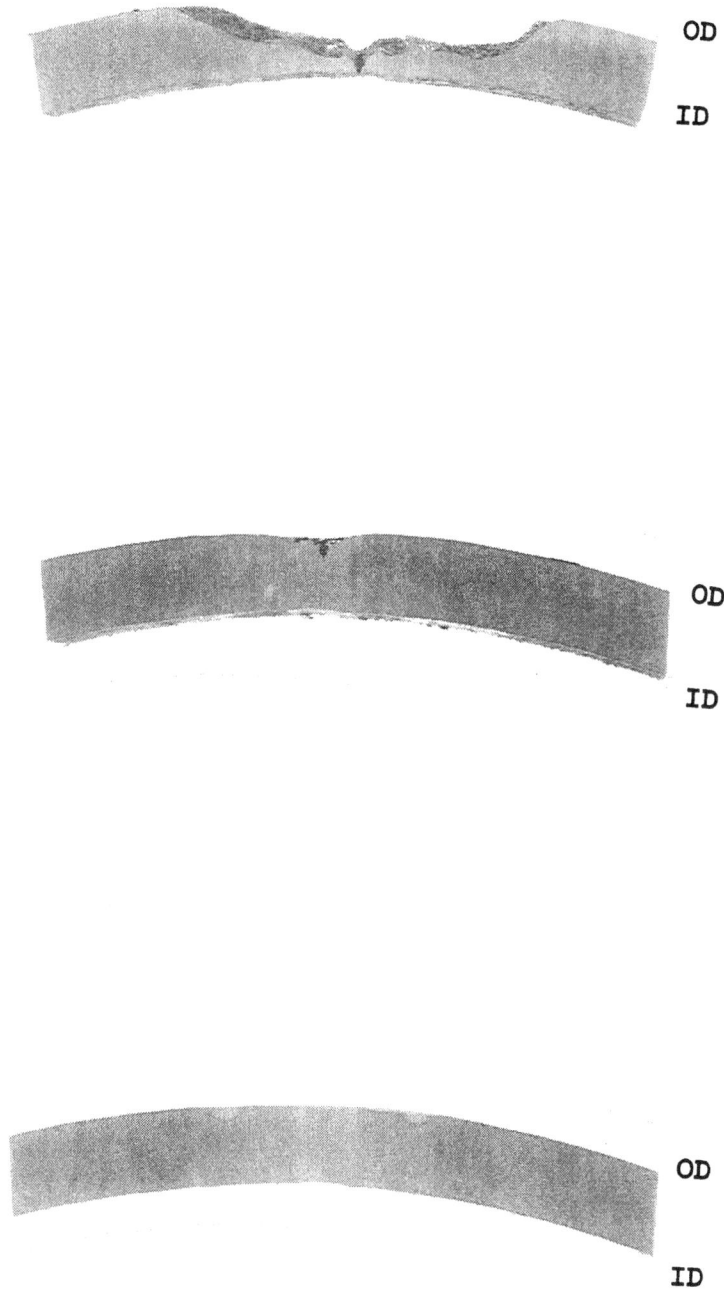

Figure 4-16.  Metallographic Sections Through Regions of Low-
Frequency ERW Seams, Two of Which Exhibit
Selective Corrosion

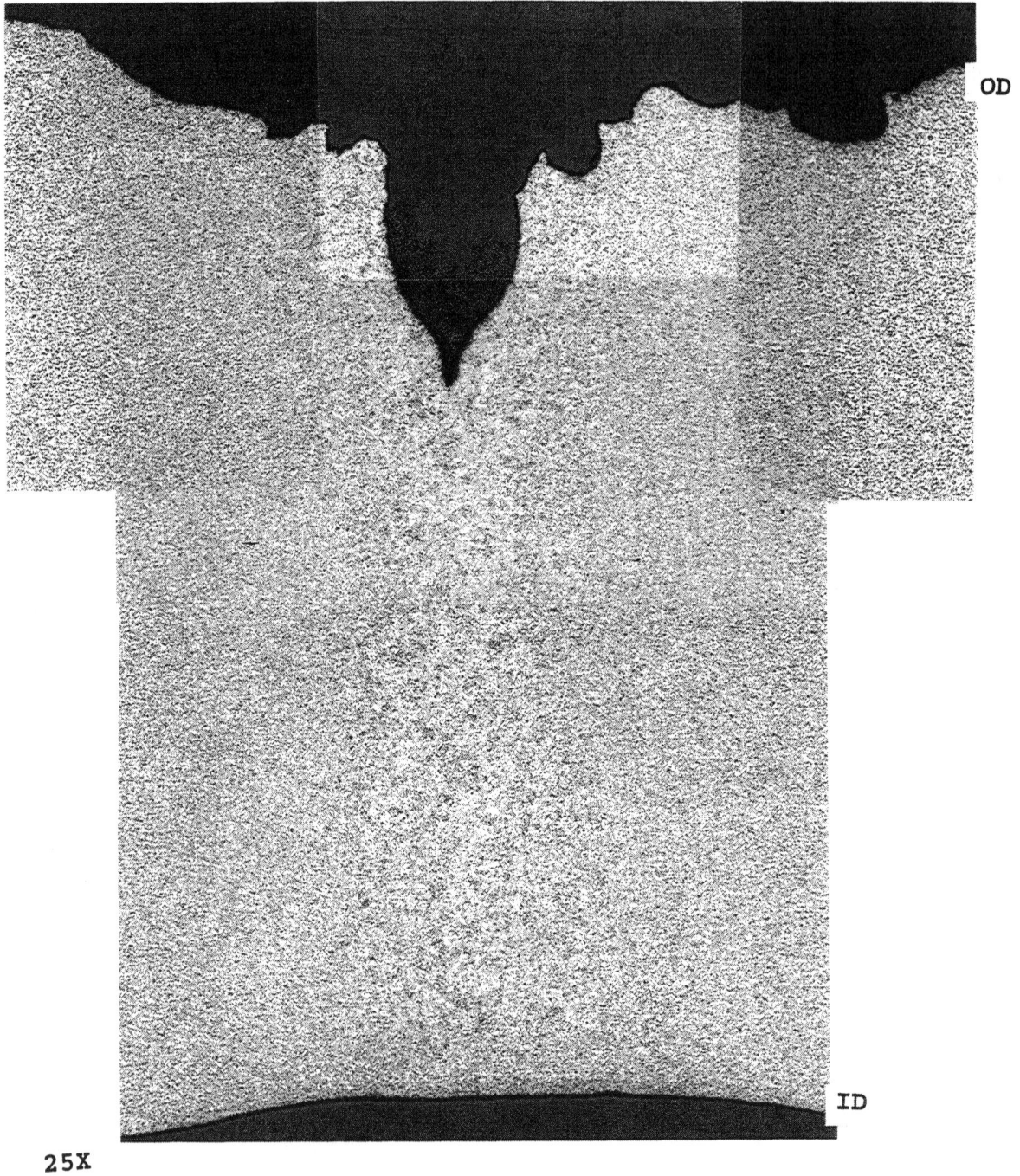

25X

Figure 4-17.    Close Up of Metallographic Section Through
Selectively Corroded ERW Seam

## 5.0 ELECTRIC-FLASH WELDED PIPE

### The Process

Electric-flash welded line pipe was made between 1930 and 1969 by only one company, A.O. Smith Corporation. A.O. Smith Corporation first produced line pipe in 1927. Their earliest pipe was made in 30-foot lengths from plates cold-formed into tubes with a longitudinal seam made from the outside by means of a shielded metal-arc process. By 1930, they were making flash-welded pipe at their Milwaukee facility in sizes ranging from 8 to 26 inches. Reference 5-1 indicates they were using "stronger" steels (not specified), making longer lengths (30 to 40 feet compared to 18 to 20 foot lengths of butt-welded and lap-welded pipe), and forming seams by a "new" welding process. Plates were pickled (descaled in an acid bath), sized, formed, welded. The pipes were then cleaned, cold-expanded, end-finished, and tested.

Reference 5-2, an A.O. Smith bulletin which appears to be about 1950-vintage, describes the process at the Milwaukee facility at that time as follows. Flash-welded pipe in sizes from 8⅝-inch OD to 26-inch OD was being made in 40-foot lengths from rolled plates. The plates were descaled by pickling, scarfed, flattened, and edge-trimmed. They were then formed into cans in three stages: U-ing, crimping, and O-ing. Lugs were then welded onto each end for positioning the can in the welder. The can was slid over a water-cooled arbor, and positioned by means of the lugs. The arbor was expanded, clamping the can against contacts near the edges. As the edges were hydraulically pressed together, the electric current was turned on to heat the edges to a semi-molten state. Just before the current was cut off, the edges were "bumped". The bumping upset the nearly molten edges, squeezing out material and forming a bond between the edges. Next, the tube was straightened. Then it was clamped in dies and flash-trimmed at both the OD and ID surfaces by moving tools leaving 1/16-inch residual flashes at both surfaces. A metallographic section across a flash-weld is shown in Figure 5-1.

At subsequent stations the ends of each length were expanded (mechanically, it appears). The entire pipe was then cold-expanded inside a horizontally divided die by means of hydrostatic pressure. The amount of circumferential stretch was said to be as much as 1⅜-inch for 26 inch pipe (corresponding to 1.7 percent expansion). The ends were then beveled or finished as required by the customer and each piece was subjected to the standard API hydrostatic test. During the test the pipe was struck at 2-foot intervals by 6½ lb hammers. Final inspection consisted of checks on diameter, roundness, and wall thickness and visual inspection of the OD and ID surfaces. The pipe was then measured for length, weighed, washed with a hot alkali solution, and stencilled. Painting with a primer was available.

The pipe was loaded onto gondola cars according to A.O. Smith's own car-loading standards.

Other information on the flash-welded pipe made between 1937 and 1943 was obtained from purchase orders for 24-inch OD and 26-inch OD pipe provided by a pipeline operator. These purchase orders illustrate that A.O. Smith made their pipe according to API standards or better, but at that time they did not have an API license. In fact, it appears that A.O. Smith Corporation did not obtain an API certificate until about 1954. However, flash welding was recognized as an approved process for the first time in the API Ninth Edition of Specification 5L (August, 1944) and in the First Edition of API Specification 5LX (February, 1948). The purchase orders also indicate that open hearth steel was to be used and that the pipe was to be tested to 90 percent of SMYS. SMYS values of 45,000 psi and 48,000 psi were included in the orders. Wall thickness tolerances of -8 and -10 percent were specified.

Through the Directory of Iron and Steel Works,[5-3] it was learned that A.O. Smith's Milwaukee facility made line pipe in sizes up to 30-inch OD starting in the mid-to-late 1950s and to 36-inch OD starting in the early 1960s. Sometime around 1970 the Milwaukee plant stopped making line pipe but continued to make flash-welded casing in the 7 to 13⅜-inch size range. A.O. Smith Corporation was bought-out by Armco in 1972.

Around 1950 A.O. Smith began making flash-welded pipe at their Houston facility. This facility was equipped with a mechanical expander, an A.O. Smith invention. Subsequently, a mechanical expander was installed at Milwaukee. The Houston facility made pipe up to 30-inch diameter in the early 1950s and up to 36-inch diameter from the mid 1950s on. In 1969 this mill was converted to a submerged-arc weld pipe mill. Their submerged arc welded pipe was unique in that the weld was started at both ends, and one pass slightly over lapped the other in the middle of the pipe length. Not too much later (about 1973) the Houston mill went out of production permanently.

Starting in 1957 A.O. Smith began continuous ultrasonic inspection of the flash-welded seams at both pipe mills. Around 1966 they began seam-normalizing the welds at both facilities using gas burners.

## Quality Factors

The quality factors for flash-welded pipe are somewhat similar to those of electric-resistance welded pipe. However, while it is probable that not every bondline was perfect (it could be susceptible to cold welds), flash-welded pipe never seemed to be plagued with the numerous bondline failures associated with some of the early ERW materials.

One of the most significant similarities between flash-welded and ERW-welded pipe is that the skelp for flash-welded pipe during the time it was made was similar in chemistry and inclusion content to the skelp typically used for ERW pipe. Hence, flash-welded pipe was susceptible to both hook cracks and selective seam corrosion. There is little doubt that the fact that flash-welded pipe was cold expanded contributed to its overall better performance than some ERW materials.

One problem which appeared in pipe made by other manufacturers but seemed to affect A.O. smith's pipe more often was that of "hard spots".[5-4] These consisted of roughly circular areas of excessive hardness in the plate having no association with the seam weld or the welding process. Because these spots occurred most frequently as multiple areas in one circumferential band, it is suspected that they were created by accidental quenching on the runout table after rolling of the plate. Perhaps they resulted from dripping water sprays which were supposed to shut off whenever the rolling process was interrupted.

## References

5-1. Graham, W. T., "Pipe Line Welding", *Natural Gas*, November, 1930, pp 28-32.

5-2. A.O. Smith Corporation, Bulletin No. 576.

5-3. Directory of Iron and Steel Works of the United States and Canada, Several Editions (1930 through 1970), American Iron and Steel Institute.

5-4. Groeneveld, T. P., and Fessler, R. R., "Hydrogen Stress Cracking Overview and Controls", 6th Symposium on Line Pipe Research, American Gas Association, Catalogue No. L30175, 1979.

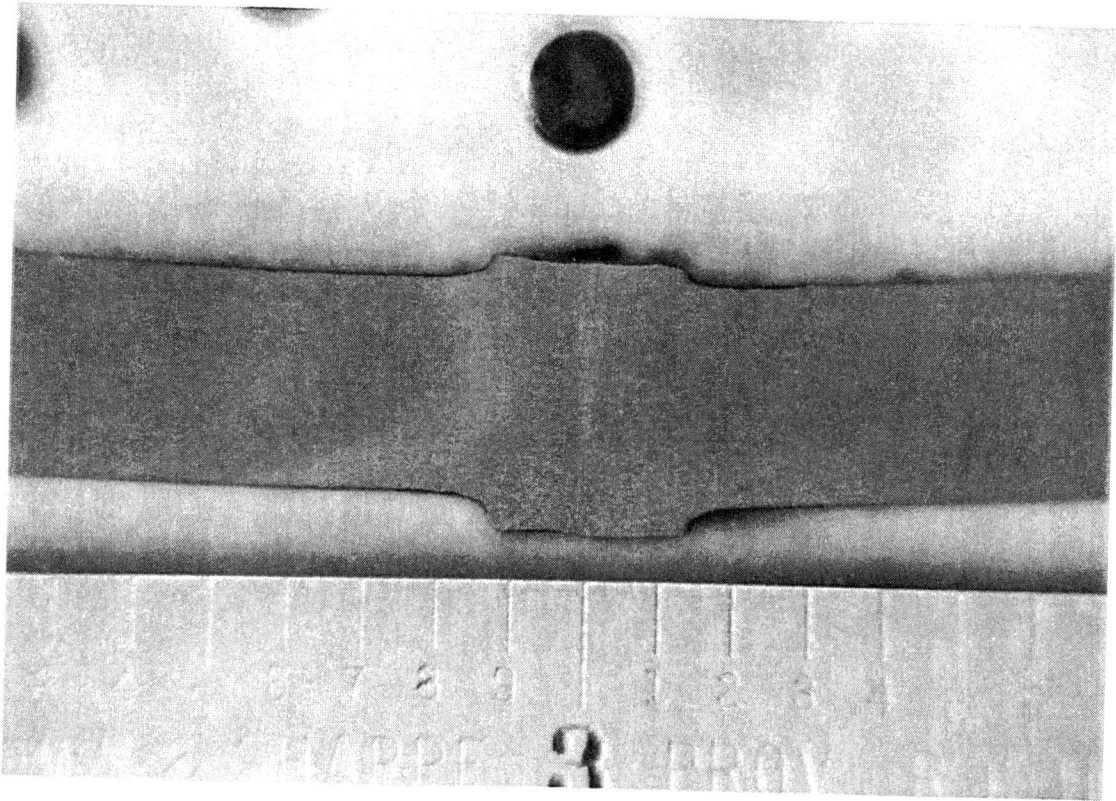

Figure 5-1.  A flash weld.

# 6.0 SUBMERGED-ARC-WELDED PIPE

Line pipe with a submerged-arc-welded seam was first made in 1930 at the Christy Park Works of National Tube Co.[6-1] The Christy Park process involved forming flat plates into "cans" in a set of presses. The edges of the cans were then joined by a single pass (multi-welder) submerged-arc weld made from the outside surface onto a backing shoe located at the ID surface. The finished pipe was rounded in a set of pyramid rolls. The Christy Park process was discontinued in 1932 for reasons unknown and some of the equipment was sold to Western Pipe and Steel Company. The Western Pipe and Steel Company became Consolidated Western Steel Company, and in 1946 they began manufacturing the first large diameter (30-inch) submerged-arc welded pipe possibly using at least some of the Christy Park equipment.[6-2] By 1948, Consolidated Western had developed a double submerged-arc-welded seam welding process and was becoming an important supplier of line pipe. A double submerged-arc-welded seam consists of at least one pass being made from the inside and one pass being made from the outside of the pipe. By 1960 several manufacturers were making double submerged-arc-welded (DSAW) line pipe,[6-3 to 6-7] and DSAW pipe remains today as one of the three principle processes for line pipe manufacturing.[6-8,6-9]

## The Process

Although varying somewhat from one manufacturer to another, the processes for making DSAW line pipe have certain common or similar steps. All such processes start with rolled plate or skelp as the raw material. Since the diameters of DSAW pipes made in North America range from 18 to 64 inches, plates are generally used instead of coiled skelp. As in the case of pipe made from coiled skelp, the rolling direction of the plate invariably ends up oriented along the axis of the pipe because of plate rolling width limitations. Any given pipe mill may receive plate from any number of sources.

At the pipe mill the plates for DSAW pipe are usually prepared by machining the ends to square them with the edges so that the edges of the formed cans will not be offset from one another. Then the edges are trimmed, planed, or beveled for welding. Some early processes planed the edges at a slight angle so that they would meet when the can was formed. Except for a small chamfer to guide the automatic welder, no bevels were made. Instead these processes depended on "burning" the weld beads into the plate. More recently, however, it is common for the edges to be machined with V-bevels from both sides with a substantial land at mid-thickness (see Figure 6-1).

Cans are cold-formed in a variety of ways. Some mills preform or "crimp" the edges as shown in Figure 6-2. Then, the plates are bent in a U-ing press (Figure 6-3) and an O-ing press (Figure 6-4). Other mills[6-9] roll-form cans in a set of "pyramid" rolls as shown in Figure 6-5. The edges of such cans are crimped after pyramid rolling to assure a high degree of roundness.

While the oldest DSAW processes involved welding from the OD surface first[6-1,6-2], all current manufactures make the inside diameter weld first usually following continuous automatic tack welding. Typically, run-on and run-off tabs are attached temporarily to the ends of the can for both the inside diameter and the outside diameter welds as shown in Figures 6-6 and 6-7 to assure uniformity from one end of the can to the other. Each of these welds is made in a single pass with multiple (2 to 5) separate wires.

Most (but not all) DSAW pipe is subjected to sizing by cold expansion after welding. Some mills use hydraulic expansion wherein internal hydrostatic pressure is used to deform the pipe against a series of rigid external dies as shown in Figure 6-8. Other mills use an internal mechanical expander comprised of segments as shown in Figure 6-9. These segments are wedged against the inside diameter surface of the pipe by a central hydraulically operated ram. As they travel radially outward to a preset limit, they expand the entire pipe successively in short segments. Berg Steel does not use either hydraulic or mechanical expansion. Instead, the welded pipes are passed through a pair of sizing rolls.[6-9]

All DSAW line pipe materials are subjected to hydrostatic testing and nondestructive testing (see Figure 6-10) of the seam and the ends are beveled as part of the final steps in the process.

## Quality Factors

While DSAW pipes may exhibit plate defects such as laminations, localized hard spots, and some of the types of defects exhibited by seamless pipe, the primary concerns with respect to the quality and serviceability of the product are associated with the seam. The following imperfections and defects can exist in DSAW line pipe. With improvements in both steel making and nondestructive testing over the years, these kinds of problems have diminished appreciably.

### Weld Metal Cracks

Weld metal cracks can arise during manufacturing from movement of the plate edges before the weld metal has cooled sufficiently. One such crack is shown in Figure 6-11. This crack occurred in the OD bead which was the first bead deposited. This was more of a problem in the early years when the techniques for holding the cans was still being perfected. In the original outside diameter-bead-first welders, the ends of every can were expected to crack because the equipment for holding the cans was ineffective at the ends. Prior to the inside diameter weld being made each such weld was back chipped and a semi-automatic "squirt" weld was used to complete the two ends of the can. A short squirt weld (about 3 inches) was sufficient at the beginning of the weld and a longer (8 inch) squirt weld was required at the end. With the advent of inside diameter-bead-first welding, better can holding equipment was developed, the practice of using run-on and run-off tabs was instituted, and the problem has largely disappeared.

## Toe Cracks

A more common and more pervasive form of cracking is the toe crack which may form at either the outside diameter or inside diameter surface where the crown of the DSAW bead intersects the plate. These cracks usually arise during cold expansion because of imperfectly round cans or excessively high weld bead crowns. Inside diameter toe cracks may form if the plate edges have a too-large radius of curvature creating a "peaked" weld. During expansion, excessive strains are concentrated at the inside diameter toes of the weld. Inside diameter toe cracks may also occur during hydraulic expansion if the outside diameter bead height is excessive and "bottoms-out" in the relief groove of the external dies. Conversely, outside diameter toe cracks can occur if the inside diameter bead is misaligned with or "bottoms-out" in the relief groove of the mechanical expander. The propensity for toe cracking is increased if the material contains large numbers of nonmetallic inclusions, particularly elongated manganese-sulfide inclusions.

One such failure from lamellar tearing of inclusions is shown in Figure 6-12. This kind of problem can be mitigated by using clean, low-sulfur steels or steels with appropriate alloying additions to effect sulfide shape control.

**Off-Seam Weld**

An off-seam weld is a condition in which the inside diameter and outside diameter beads are offset circumferentially from one another. The condition is acceptable unless a void exists because the two beads do not meet. The weld shown in Figure 6-11 was an off-seam weld which should have been rejected because the ID bead did not fuse with the plate on one side. It caused a part-through crack in the outside diameter weld bead which led to a hydrostatic test failure.

**Offset Plate Edges**

Occasionally, if the cans are cambered or twisted or have wavy edges, the two edges will end up being radially offset from one another. This condition results in a bending stress being associated with internal pressure. While the added stress is not particularly harmful under static load, it can lead to a shortened time to failure if significant pressure cycles are applied and a fatigue-crack initiator exists (see rail shipment fatigue below).

**Welding Defects**

As with all welds involving filler metal, imperfections or defects can arise from porosity, slag inclusions, inadequate penetration, inadequate fusion and undercut. These kinds of problems can be avoided by good quality control on the part of the manufacturer.

**Transit Fatigue**

Transit fatigue refers to cracks which form in pipes as they are being transported by rail, truck, or ship.[6-10] The cracks are created by repeated stresses from shaking or bouncing of the load from the motion of the transporting vehicle. Typically, the most likely pipes to be affected will be those on or near the bottom of the stack. The nature of the bearing surface is an important factor. Concentrated bearing loadings tend to create localized deformations and/or contact damage in the form of dents or abrasion. Such areas have served as crack initiators. Fatigue cracks arising in this manner have been large enough to cause hydrostatic test failures. But, the more insidious cases involve those which are too small to fail in the initial hydrostatic test of the pipeline and survive to become enlarged by service pressure cycles to the point where they produce a service failure.

Both seamless and ERW pipe materials have been affected by transit fatigue, but the more widely known cases have occurred in DSAW line pipe usually associated with rail shipment fatigue cracks. The problem with DSAW pipe typically arises if and when large D/t pipes are loaded with the seam weld bearing on a bearing strip, especially if the pipe is on the bottom row. When shipped in this manner, high-cycle, low amplitude vibrational stresses can initiate cracks at the toes of the weld. Shipment cracks may form at the toe of the weld on the outside diameter surface or the inside diameter surface or at both surfaces. When this has occurred in the past, the cracks have frequently been too small to cause failure during the field hydrostatic test.

When subjected to large numbers of significant pressure cycles in service (as in some liquid pipelines), these cracks have been known to grow to failure. The majority of those which have failed have also been found to be associated with offset plate edges. Thus, it is apparent that the fatigue life tends to be shorter because of the added bending stress. An example of a rail shipment crack is shown in Figure 6-13. This one caused a hydrostatic test failure. Note the degree of offset of the plate edges. The best means of avoiding transit fatigue is to assure that loads of pipe are properly stacked with adequate bearing and separating strips made from wood. API has issued recommended practices (API RP 5L1 for rail car loading and API RP 5W for ship or barge loading). By having pipe materials loaded according to these practices, a purchaser can significantly reduce the exposure to transit fatigue.

## Quality Control

The kinds of problems alluded to above can be minimized or avoided by quality control on the part of the manufacturer and by adequate diligence on the part of the purchaser. Visual and nondestructive inspections and hydrostatic testing at the mill are mandated by the API 5L Specification and can be supplemented by negotiation. While the manufacturer is responsible for the quality of the product, there is no question that an informed and diligent purchaser will tend to receive a better product than one who takes a relatively passive approach. The nondestructive and visual inspections that are mandated should be able to detect any welding problems, and permit the purchaser to reject any unacceptable pipes. Finally, rail shipment fatigue can be avoided by proper rail car loading, but a purchaser must provide

for the loading to be done by API Recommended Practice 5L1 and
assure that the practice is rigorously observed.

## Material Factors

As in the case of ERW line pipe the use of a flat hot-
rolled material allows the purchaser of DSAW line pipe an
excellent opportunity to obtain a fracture resistant, highly
weldable material. Through current steel making and continuous
casting technology and the use of thermomechanical means (i.e.,
controlled-rolling), a manufacturer can achieve both high
toughness and good weldability. The limiting of sulfur and
carbon to the lowest possible levels leads to clean (low
inclusion), low-transition-temperature, high-ductile-toughness
line pipe. The tendency toward lower transition temperatures is
enhanced by controlled rolling which produces smaller grain
sizes. The reduction of carbon levels leads to improved
weldability. The purchaser can negotiate specific toughness and
transition temperature levels by utilizing Supplementary
Requirement SR5 in the API 5L Specification. Weldability can be
assured by specifying an appropriate limit on "carbon
equivalent".

## References

6-1.    Anderson, G. C., "Design and Operation of Electric Weld
        Pipe Mill at National Tube's National Works", *Iron and
        Steel Engineer*, May 1952, pp 96-101.

6-2.    Klick, M. P., "Pipe Mill, Utah", *Iron and Steel
        Engineer*, August 1956, pp 111-117.

6-3.    "Electric Weld Tube Mill Goes into Production at
        National Tube Co's McKeesport Works, *Iron and Steel
        Engineer*, June 1950, pp 120-122.

6-4.    Brecher, F., "Steelton Facilities for Manufacturing Expanded Line Pipe", *Iron and Steel Engineer*, May 1962, pp 104-109.

6-5.    Jonas, E. A., "Quality Control Features in the Manufacture of Submerged-Arc-Welded Pipe", *Mechanical Working of Steel 1*, AIME, 1963.

6-6.    Middleton, J. M., "Manufacture of Electric Fusion Welded Pipe", *Iron and Steel Engineer*, March 1951, pp 66-69

6-7.    "Kaiser Doubles Napa's Pipemaking Capacity", *Iron and Steel Engineer*, June 1959, p 153.

6-8.    Labee, C. J., "Baytown Revisited--U.S. Steel Texas Works' Update", *Iron and Steel Engineer*, March 1978, pp T-1, T-12.

6-9.    Wise, R. L., and Hodgson, A. W., "Berg Pipe: A Mill Whose Time Has Arrived", *Iron and Steel Engineer*, August 1983, pp 36-40.

6-10.   Bruno, T. V., "How to Prevent Transit Fatigue to Tubular Goods", *Pipe Line Industry*, July 1988.

Figure 6-1.  Plates with Edges Beveled for Welding

*[Berg Steel Pipe Corp.]*

Figure 6-2.  Edge Crimping

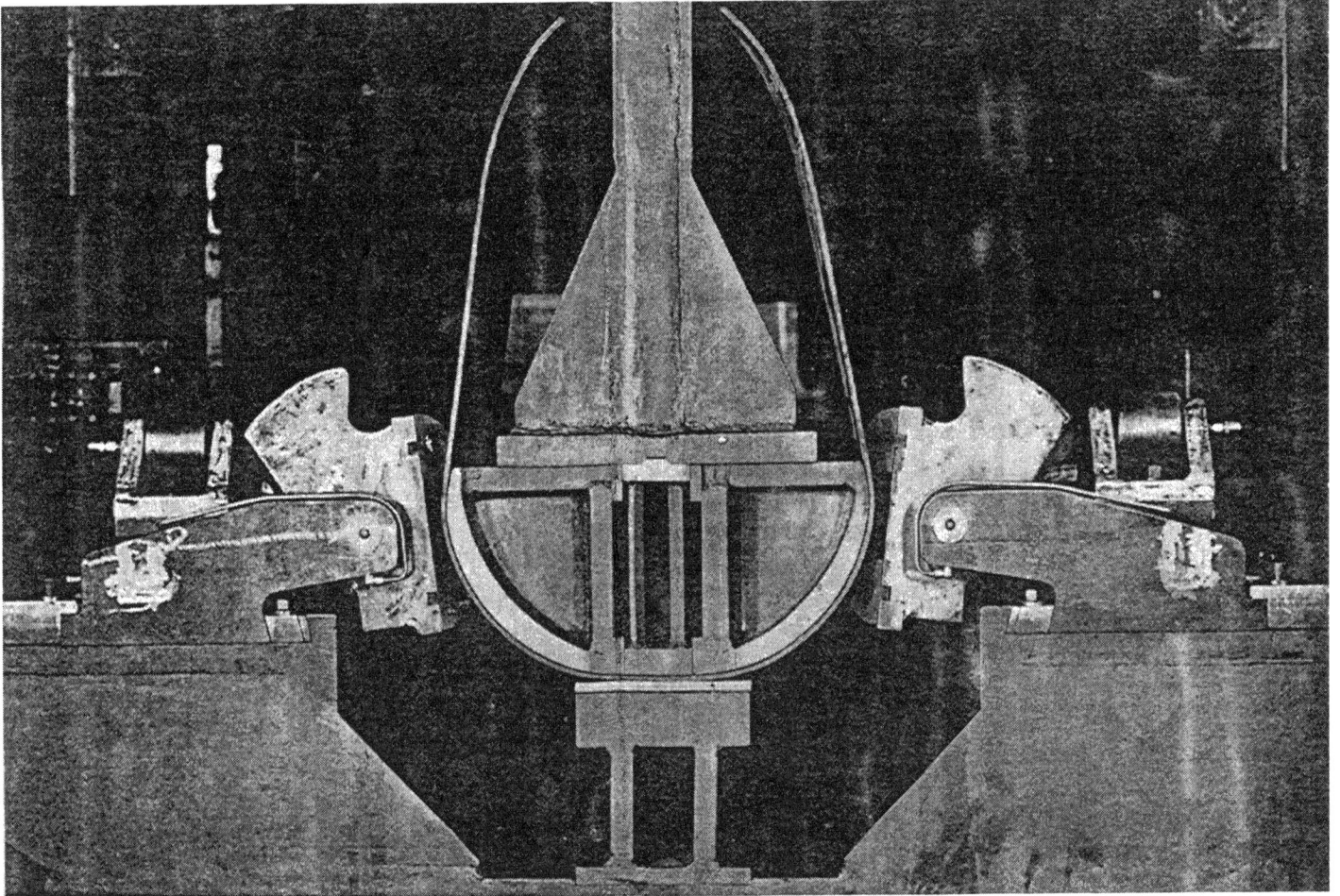

Figure 6-3.  U-ing

Figure 6-4.  O-ing

Figure 6-5. Pyramid Rolls

*[Reprinted with permission from The Making Shipping and Treating of Steel, copyright 1985, Association of Iron and Steel Engineers]*

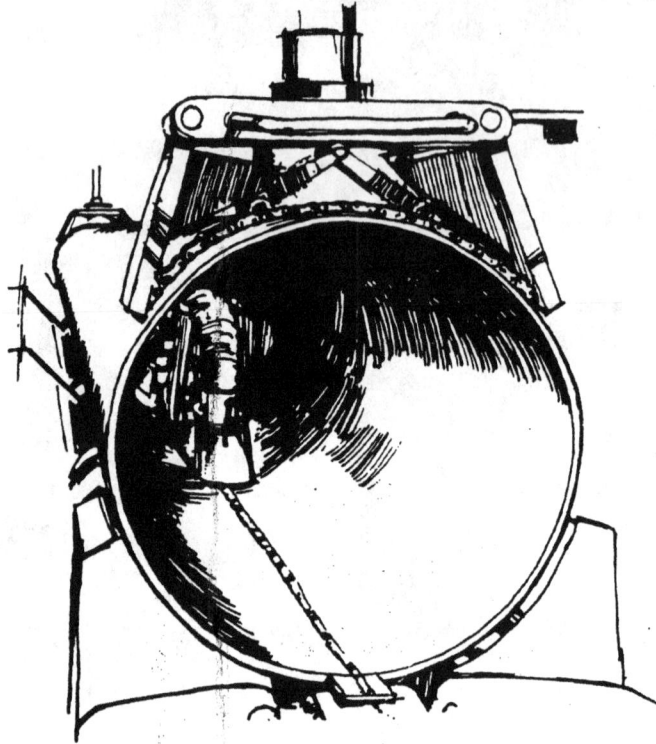

Figure 6-6.    Inside Welding

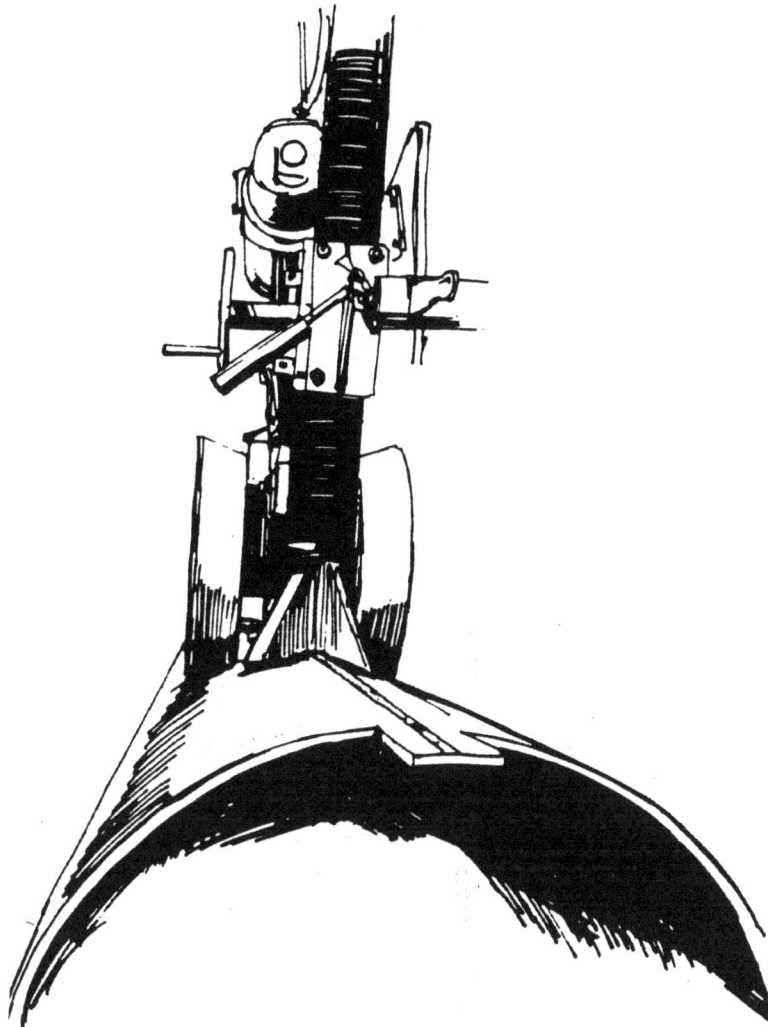

Figure 6-7.   Outside Welding

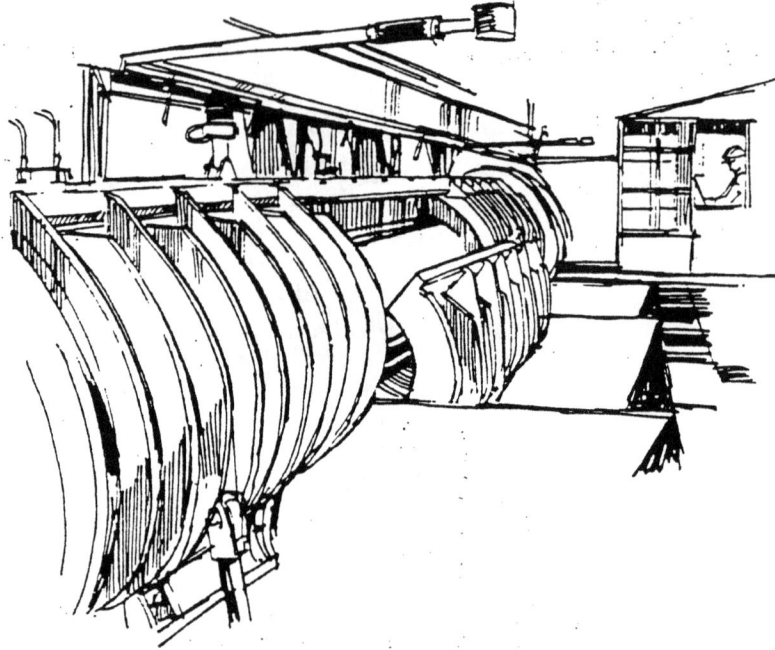

Figure 6-8.   Hydraulic Expander

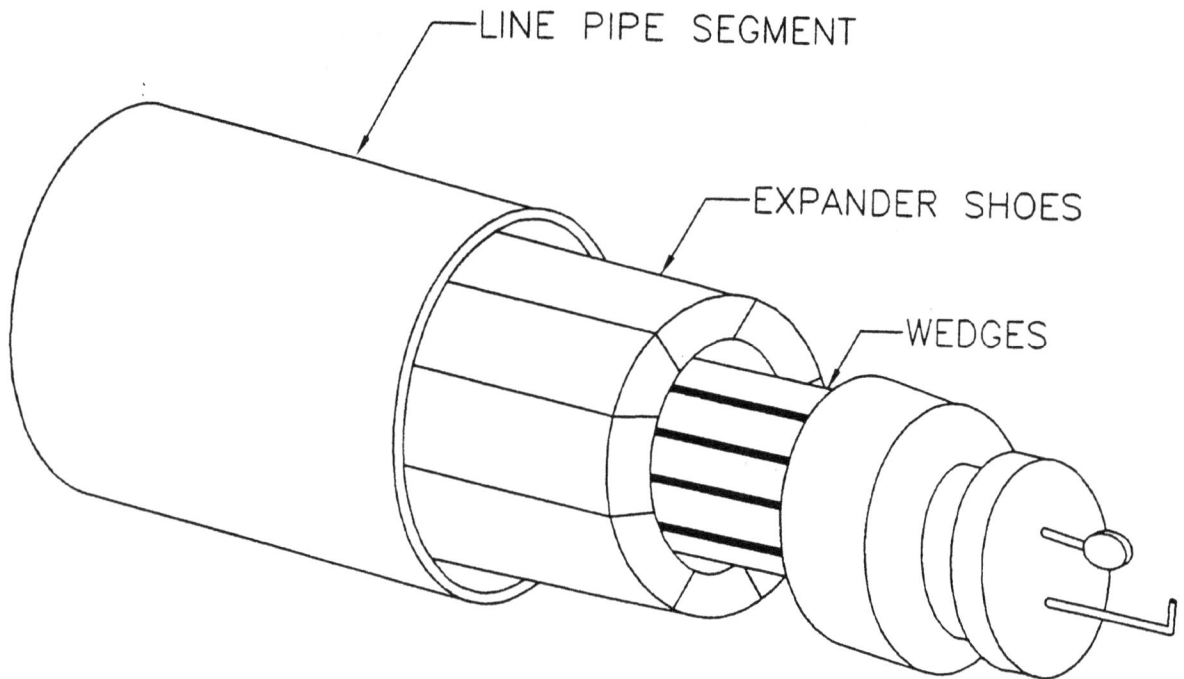

Figure 6-9.   Mechanical Expander

*[Courtesy of Napa Pipe Corp.]*

Figure 6-10. UT Inspection

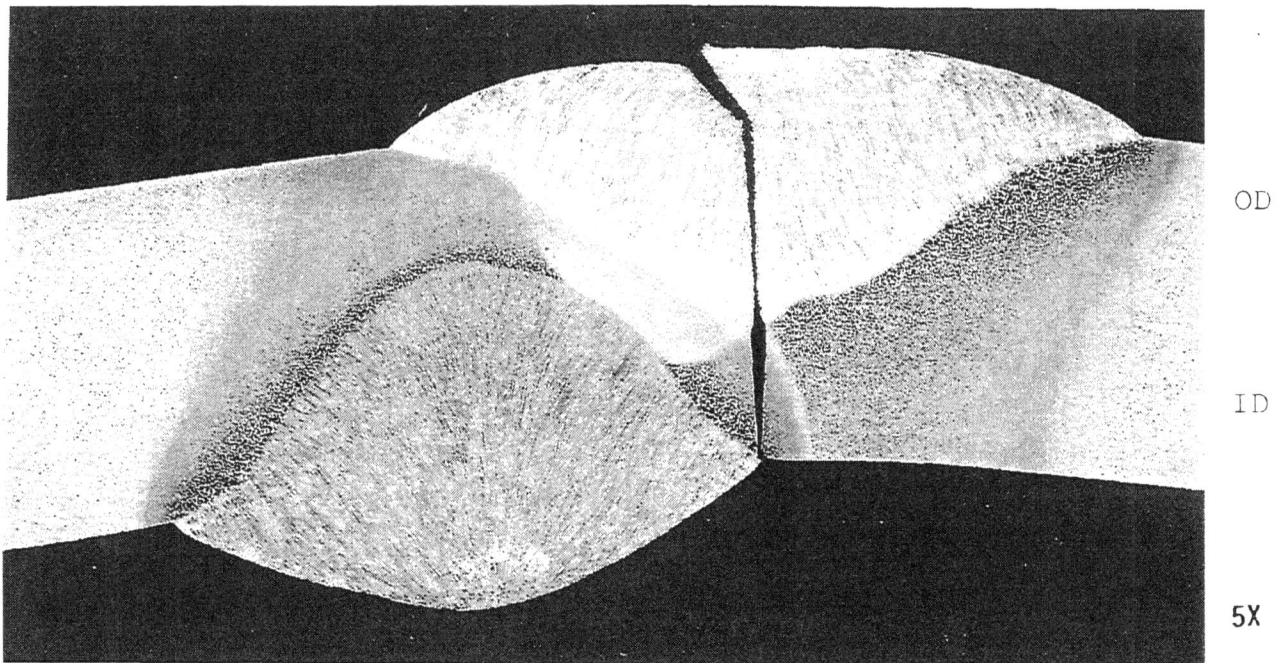

OD

ID

5X

Figure 6-11. Hydrostatic Test Failure Caused by Offseam Weld and Weld Metal Crack

**10X**

Figure 6-12.  Lamelar Tearing at Toe of DSAW Seam (Note That
Wall Thickness Had Been Reduced by Grinding)

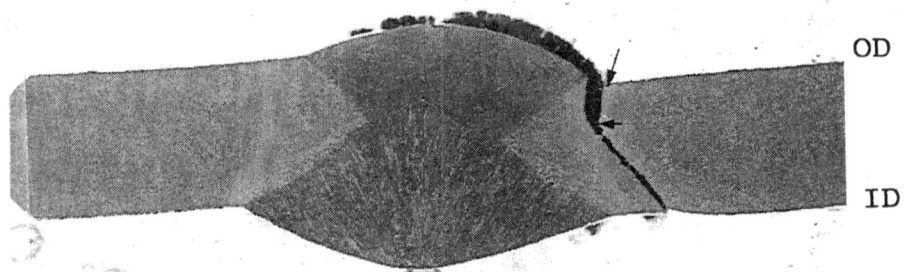

**2.8X**

Figure 6-13.  Rail Shipment Crack Extending from OD Toe of DSAW
Seam.  The Rest of the Crack is the Ductile
Rupture Which Occurred During a Hydrostatic Test

# 7.0 SPIRAL-WELD LINE PIPE

## Historical Perspective

The available literature suggests that spiral-weld pipe was first produced in the U.S. in 1888 using a hammer lap-welding process.[7-1,7-2] The production speed of such pipe was said to be 1 foot per minute and diameters from 4 to 30 inches were produced. Strips of steel or iron skelp were helically wound, overlapped and heated at the edges and hammer-welded. A sketch of this original spiral welder is shown in Figure 7-1.

Apparently, other manufacturers developed similar processes for making spiral-weld pipe that were used to make pipe according to ASTM specifications (A139 and A211).[7-1] By 1960 at least one U.S. manufacturer was able to make electric-resistance welded spiral-seam pipe in sizes 6⅝ to 12¾ inches meeting ASTM Specification A135.[7-3] (See Figure 7-2). There is evidence that at least one manufacturer attempted to make spiral-weld pipe by both electric-resistance butt-welding and by electric-resistance lap welding.[7-4]

By 1965, at least two German manufacturers and several North American mills including Lone Star, Acme Newport, and Taylor Forge had developed the capability to produce spiral-weld pipe with a double-submerged-arc welded seam.[7-1,7-2,7-3] Since the API specifications for line pipe (5L and 5LX) at the time accepted welded pipe with a longitudinal seam only, a tentative Specification 5LS was developed to permit the use of spiral-weld pipe for line pipe. In 1967, the tentative specification became a standard (API STD 5LS). "Prior to 1967, the design codes for liquid petroleum and natural gas pipelines in the U.S. penalized spiral seam pipe with a low joint factor allowance. Since there was no penalty under the American Water Works Association code, much of the early development of spiral weld pipe was for water lines. Operating stresses were lower than for oil and gas and so weld quality tended to be of marginal importance. European codes did not penalize the joint factor and with the advent of API Specification 5LS the U.S. codes changed to the same factor as DSAW straight seam. With the exception of Canada, the North American market for spiral-weld pipe never really developed".[7-3]

When the 5LS standard was first issued, it permitted the seam to be made either by double submerged-arc welding or electric-resistance welding. In 1983 when the Specifications 5L, 5LX, and 5LS were combined into one Specification 5L, only the double-submerged-arc process was approved for making spiral-weld pipe. Although numerous North American pipe manufacturers have made and continue to make spiral-weld pipe, only three (Interprovincial Steel and Pipe Corporation, Ltd., Lone Star Steel Company,[7-5] and Stelco Inc.) held certificates to produce API Specification 5LS line pipe.

Spiral pipe mills tended to required lower capital investment than straight-seam (u and o) mills but they also tended to have lower rates of productivity.[7-3]

## The Process

A schematic view of the process is shown in Figure 7-3. The arrangement shown is used by a manufacturer of submerged-arc welded spiral-seam pipe. Coils or plates may be used as feedstock. If coils are used, the coil is uncoiled, leveled, rotary-edge-sheared, and skived in preparation for forming and welding. Continuous operation involving welding one coil to the next is possible because the uncoiler and leveler unit is mounted on rails and can be moved along with the skelp long enough to permit skelp welding.

The skelp is formed by a cage-roll forming unit. In some mills the edges of the skelp are joined by a continuous tack weld which is subsequently remelted by the ID and/or OD beads.

In other mills the ID weld is deposited immediately, followed by the OD bead. Both beads are deposited by automatic multi-wire submerged-arc welders.

After the welding has been completed and the weld has cooled appropriately, the seam is continuously inspected by an ultrasonic unit. A flying cut-off machine cuts individual lengths of pipe which are then sent for further processing including, at a minimum, end preparation, hydrostatic testing and final inspection. At least one mill (Stelco) sizes the pipe by mechanical expansion. The expander shoes have grooves appropriately located to accommodate the spiral seam. Also, different mills may treat the skelp end weld differently. In most cases, a single submerged-arc-weld pass is made initially from one side to join the skelp. This becomes the ID bead when the pipe is formed. The OD bead of the skelp end weld is completed at a later stage. The junctions of skelp end welds with the spiral seam are called T-welds, and they are subjected to film radiographic inspection.

The versatility of spiral-weld pipe is illustrated in Figure 7-4. The angle $\alpha$ can be varied over a fairly wide range so that a wide range of diameters is possible. Using wide skelp and a small angle, a manufacturer can make quite large pipe. It is noted that the API specification limits the B to D ratio (skelp width-to-pipe diameter ratio) to a range of 0.8 to 3.0. Correspondingly, $\alpha$ may range from 14.8 to 72.7 degrees.

## Quality Factors

The orientation of the skelp rolling direction in spiral-weld pipe gives it a unique advantage over longitudinal-seam pipe. It is well-known that rolling creates anisotropic properties, and in particular it results in higher resistance to ductile crack propagation across the rolling direction than along it. The difference appears to range from a factor of 1.5 to a factor of 3. Thus, in a longitudinal-seam pipe, the greatest fracture resistance is orientated with respect to a circumferential crack and the least fracture resistance is oriented with respect to an axial crack. Unfortunately, axial cracks and axial crack propagation are by far the greatest concern. Spiral-weld pipe, in contrast to longitudinal-seam pipe, orients the anisotropy with respect to the angle $\alpha$ shown in Figure 7-4. If $\alpha$ is 45 degrees or less, the material will exhibit much better axial fracture resistance than it would if made into a straight-seam pipe. Reference 7-2 presents data which confirms that at least a 40 percent increase in resistance to axial crack propagation was exhibited by one material because of its $\alpha$ angle of 22.5 degrees. This kind of advantage could prove to be important for pipelines where ductile fracture propagation is a significant risk.

As is the case with straight-seam DSAW pipe, spiral weld DSAW pipe is subject to the type of weld defects that may be present in any DSAW weld.

## References

7-1. Groh, H., "Manufacture of Spiral Weld Pipe," *Journal of Metals*, October, 1965, pp 1141-1148.

7-2. Sommer, B., "Spiral-Weld Pipe Meets High Pressure Needs," *Oil and Gas Journal*, February 1, 1982, pp 106-116.

7-3. Von Rosenberg, E. L., private communication, (1996).

7-4. Taylor Forge Bulletin 599.

7-5. Heald, T.S., "Radio-Frequency Resistance Welding of Carbon Steel Pipe and Tubing," *Mechanical Working of Steel 1*, AIME, 963, pp 23-31.

7-5. Welch, D.E., "Recent Developments in the Manufacture of Tubular Products at Lone Star Steel," *Iron and Steel Engineer*, August, 1963, pp 113-115.

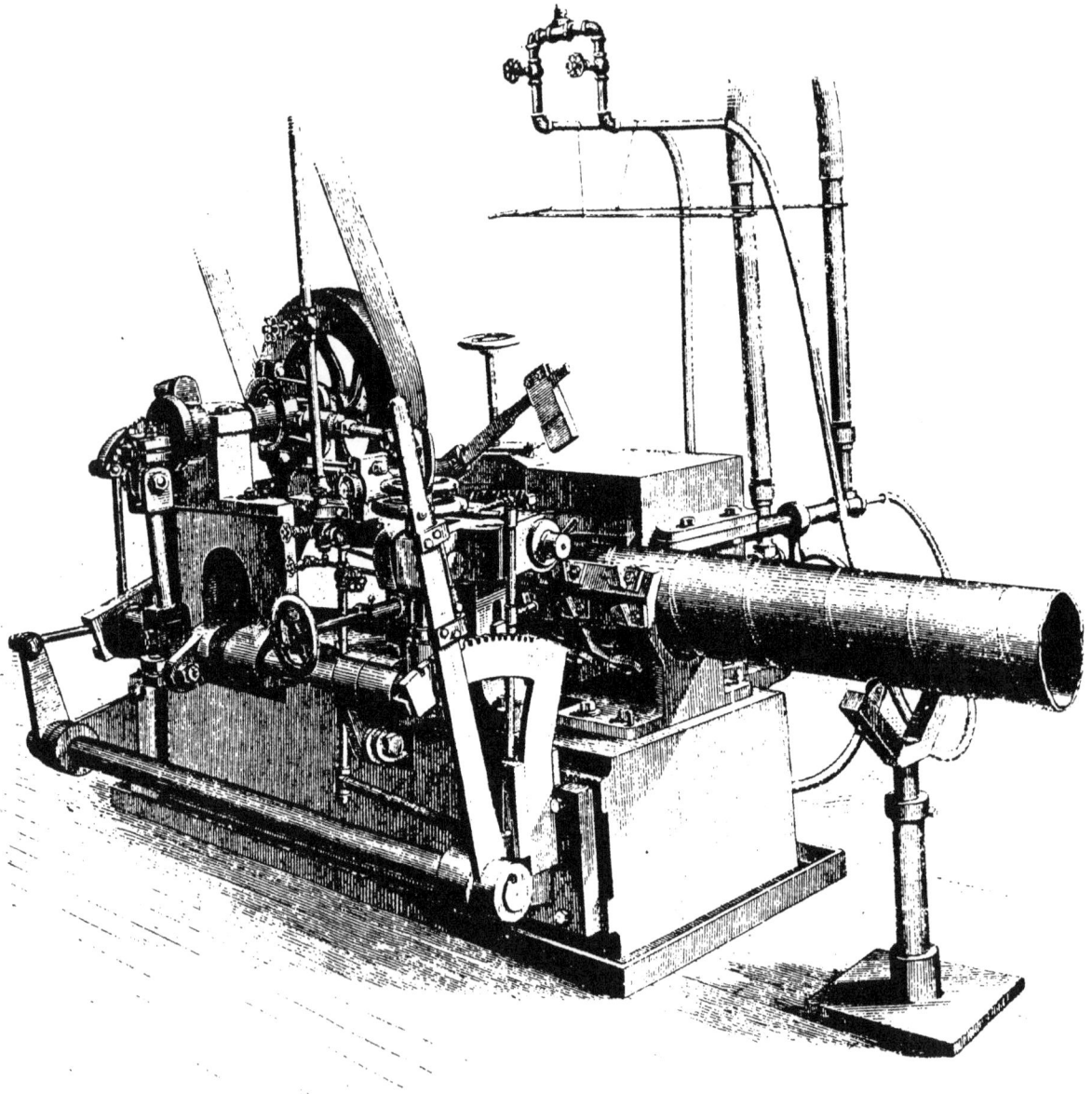

MACHINE FOR THE MANUFACTURE OF SPIRALLY WELDED TUBING, AT THE WORKS OF THE SPIRAL WELD TUBE COMPANY, EAST ORANGE, N. J.

[MHP Mannesmann Hoesch Präzisrohr GmbH]

Figure 7-1.    Early Spiral-Seam Hammer Lap Welder.

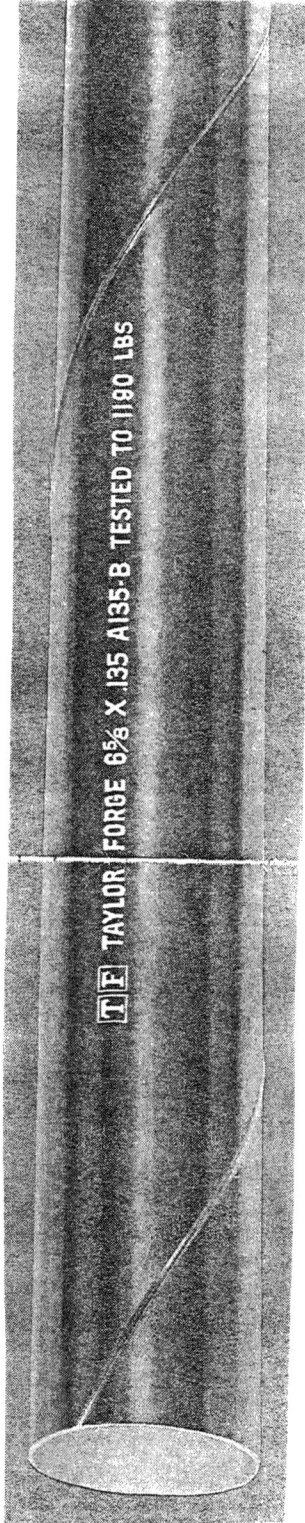

*[Taylor Forge International, Inc.]*

Figure 7-2.   1960-Vintage Spiral-Seam Electric-Resistance Welded Pipe.

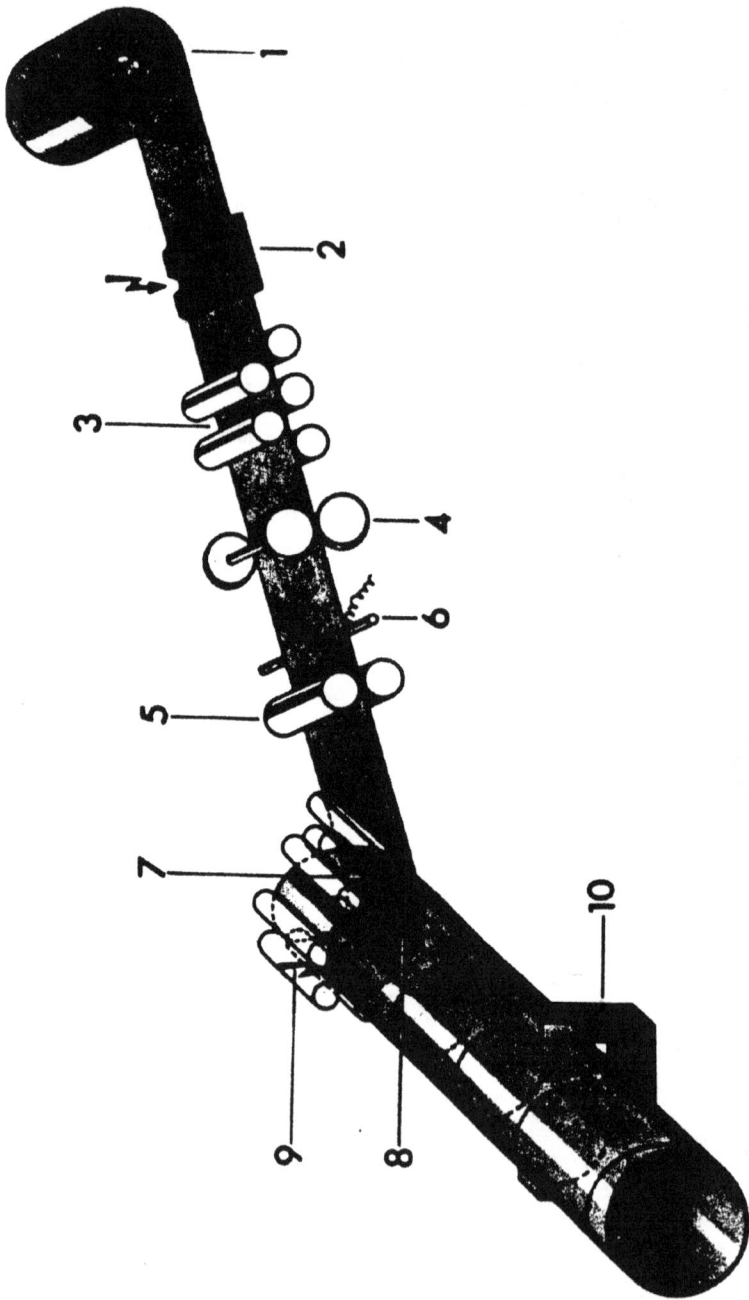

1. Coil
2. Welding unit for joining of coil ends
3. Roller leveller
4. Circular edge cutters
5. Feed rolls
6. Planing tools
7. Forming unit
8. Inside welding
9. Outside welding
10. Diameter control

[MHP Mannesmann Hoesch Präzisrohr GmbH]

Figure 7-3.    Diagram of Spiral-Weld Production.

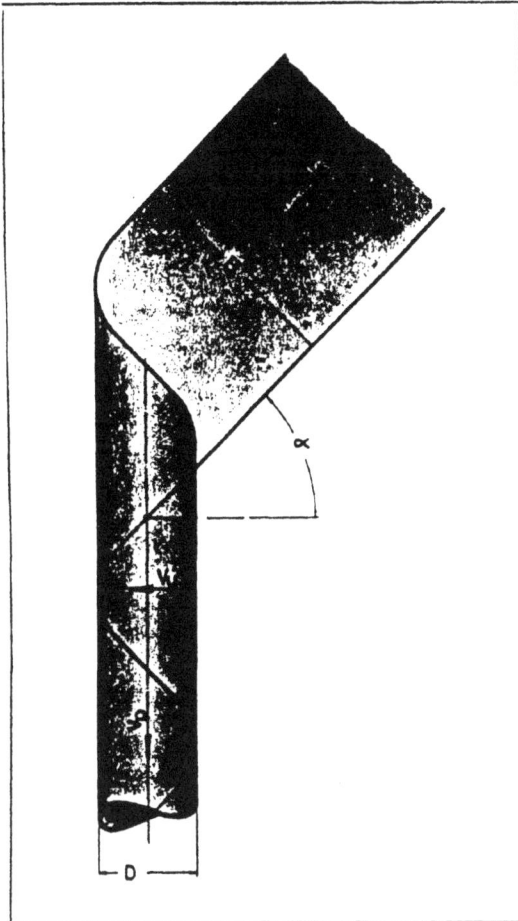

Fig. 3

Formula Symbols

D   = Pipe Diameter

B   = Width of strip

$V_S$ = Welding speed

$\alpha$   = Running-in angle of the strip

$V_P$ = Production Speed

S   = Length of spiral seam per meter of pipe

$V_U$ = Peripheral speed

$$S = \frac{D \cdot \pi}{B}$$

$$\sin \alpha = \frac{B}{D \cdot \pi}$$

$$V_P = V_S \cdot \sin \alpha$$

$$V_U = V_S \cdot \cos \alpha$$

*[MHP Mannesmann Hoesch Präzisrohr GmbH]*

For manufacturing a pipe with a diameter "D" using a strip with the width "B", the machine base frame is set to the corresponding running-in angle $\alpha$.

Figure 7-4.    Forming Principle for the Manufacture of Spiral-Weld Pipe.

# SECTION C

## MANUFACTURERS OF LINE PIPE IN NORTH AMERICA

Presented in this section are tables of line pipe manufacturers grouped by pipe manufacturing processes. Because of the enormous number of manufacturers these lists are limited (in most cases) to manufacturers who made or are making API line pipe. The information available differs from one table to the next because in some cases, especially for the older processes, limited information was found. The tables are as follows:

| TABLE | TITLE |
|-------|-------|
| C-1 | Past Manufacturers of Butt-Welded or Continuous Welded Pipe |
| C-2 | Current Manufacturers of Continuous-Welded Pipe |
| C-3 | Manufacturers of Lap-Welded Pipe |
| C-4 | Seamless Pipe Facilities Past and Present |
| C-5 | Past Manufacturers of ERW Pipe |
| C-6 | Current Manufacturers of ERW Pipe |
| C-7 | Past Manufacturers of Submerged-Arc Welded Pipe |
| C-8 | Current Manufacturers of Submerged-Arc Welded Pipe |
| C-9 | Current and Past Manufacturers of Spiral-Welded Pipe |

Tables C-1 and C-3 cover manufacturers of butt-welded, continuous-welded, and lap-welded pipe which are no longer producing line pipe made by these methods. The tables present key dates, sizes manufactured, and other information if available. Table C-2 presents the manufacturers which are currently making continuous-welded pipe. Table C-4 presents the current and past manufacturers of seamless pipe. The processes used by the various manufacturers are described. Other information includes mill capacities if available, key dates, diameters manufactured, and inspection and quality factors.

The ERW pipe manufacturers are divided into two tables, Table C-5 for manufacturers which either no longer are in business or have closed particular ERW facilities. Table C-6 presents the manufacturers currently making ERW pipe. In both tables the processes used by each manufacturer are described. Other information includes mill capacities, key dates, whether or not the pipe was cold expanded, the hydrostatic test pressure limit, the sizes and grades manufactured, and the inspection and quality factors.

The manufacturers of straight-seam submerged-arc welded line pipe are presented in two tables (C-7 and C-8) on the same basis and with the same types of information as that presented for ERW pipe manufacturers.

Table C-9 presents the information on current and past manufacturers of spiral-weld line pipe. The information included is the same as that given in Tables C-5 through C-8.

No table is presented for flash-welded line pipe because only one manufacturer, A.O. Smith Corporation, made flash-welded pipe. The processes used by A.O. Smith and other pertinent information are described in the section on flash-welded pipe.

Most of the information in these tables comes from obsolete or current brochures obtained directly or indirectly from the manufacturers. Additional information was obtained from articles in the open literature referenced in other sections of this document, from private sources, and from several editions of the "Directory of Iron and Steel Works of the United States and Canada", American Iron and Steel Institute.

The following manufacturers do not appear in any of the tables even though they are known to have held API 5L and/or 5LX Licenses. The reason they do not appear in any table is that we were unable to obtain any information regarding their capabilities.

|  | 5L | 5LX |
|---|---|---|
| American Steel Export Co., New York, NY | 1930, 1931 | — |
| Cherokee Steel Co., Tulsa, OK | — | 1962-1965 |
| Donovan Steel Tube Co., Toledo, OH | 1982, 1983 | 1982 |
| Fort Worth Pipe & Supply, Fort Worth, TX | 1980-1987 | 1980-1982 |
| Harrisburg Steel Corp., Harrisburg, PA | 1940-1954 | — |
| Ingenieria Mecanica Tubular SA de CV, Tlalnepantla, Mexico | 1995 | — |
| Mario Maraldi SpA, Canada | — | 1968-1970 |
| Maruichi America Corp., Santa Fe Springs, CA | 1980-1987 | 1980-1982 |
| Master Tank & Welding, Dallas, TX | — | 1951-1965 |
| McNamar Boiler & Tank Co., Tulsa, OK | — | 1951-1958 |
| Midwest Specialty, Tulsa, OK | 1987 | — |
| NAPSCO, Bensalem, PA | 1983-1987 | — |
| National Annealing Box Co., Washington, PA | 1983-1984 | — |
| Palmer Tube Mills, Inc., Chicago, IL | 1995 | — |
| Paragon Industries Inc., Sapulpa, OK | 1995 | — |
| R.O. Industries, Newton Falls, OH | 1984, 1985 | — |
| Smith-Scott Co., Riverside, CA | 1962-1968 | 1962-1971 |
| Sonco Steel Tube Inc., Brampton, Ont. | 1983-1987 | — |
| Steel Forgings, Inc., Shreveport, LA | 1995 | — |
| Thompson Pipe & Steel Co., Denver, CO | 1985-1987 | — |
| TransAmerica Pipe Corp., Bensalem, PA | 1982 | 1982 |
| Trinity Industries, Inc., Dallas, TX | — | 1966-1973 |
| Tuberia y Estructuras SA, Mexico, DF | 1983, 1984 | — |
| United Concrete Pipe Corp., Baldwin Park, CA | 1969, 1970 | 1969, 1970 |

# Table C-1. Past Manufacturers of Butt-Welded or Continuous-Welded Pipe

| Manufacturer | Mill/ Capacity | Type Butt or Continuous | Key Dates | Diameter Range, inch | Std API Threaded Pipe | Plain End Line Pipe | Other Information | Historical Information and Predecessor Companies |
|---|---|---|---|---|---|---|---|---|
| Bethlehem Steel Corp | Sparrows Point, MD 2 mills 75,000-360,000 t/yr | BW | Began Operating in 1930 to 1940 | ½ to 4 | | | | Replaced by CW mills in 1940. 1930 through 1938 was Maryland Plant at Sparrows Point |
| | Sparrows Point, MD 2 mills | CW | Began operating in 1940 | ½ to 1½ and 2 to 4 | | | | |
| A.M. Byers Company | South Side Works, Pittsburgh, PA 47,000 t/yr | BW | 1938 through 1964 | ¼ to 4 | Wrought iron and steel | Wrought iron only | Marked iron pipe with red paint spiral to distinguish from steel | Established as wrought iron producer in 1864. Rolled "Byers Iron" or "Byers Steel" marking on pipe. |
| | Pittsburgh, PA 65,000 t/yr, 1 mill | BW | 1920 through 1935 | ¼ to 4 | | | | |
| Central Tube Company | Pittsburgh, PA, 4 furnaces 236,000 t/yr | | 1935 through 1938 | ¼ to 3 | | | | |
| Dominion Steel and Coal Co. | Montreal, Quebec, Canada | CW | Began 1960. No longer producing line pipe. | ¾ to 4 | | | | |
| Jones & Laughlin Steel Corp | Aliquippa Works, Woodland, PA, 2 mills, 1926, 1930, 4 mills | BW | 1926 to 1957 | ¼ to 3 | | | | |
| | Aliquippa, Works, Aliquippa, PA, 2 mills, 120,000 to 420,000 t/yr | CW | 1957 to closure of Aliquippa Works | ½ to 4 | | | | |
| Kaiser Steel Corp | Works at Fontanta, CA, 144,000 t/yr in 1948, 135,000 t/yr, 1960 120,000 t/yr, | CW | | ½ to 4 from 1951 to 1957 ½ to 4 I.D.(1960 to 1970) | | | Grade A25 | Kaiser Co, Inc., 1948 |
| National Supply Company | Spang Works, Etna, PA, 2 mills, (1926-1938 had 3 mills) 87,000 t/yr | BW | 1920 to 1938 | ⅛ to 3 | | | | Spang, Chalfant and Company, Inc. 1920 to October 23, 1937(Subsidiary of Armco Steel Corp.), 1948 to 1960 know as Works at Etna, PA also known as Etna Iron & Tube Works |
| | Spang Works, Etna, PA, 2 mills, 250,000 t/yr | CW | 1939 to 1960 | 0.840 to 4½ | | | 75 to 300 fpm, flying not SAW, coil end flash welded, 1000 psi test | |
| National Tube Co., Inc | Continential Works, Pittsburgh, PA, 2 mills, 92,000 t/yr | BW | 1920 (last known year of operation) | ¼ to 2 | | | | Now a subsidiary of U.S. Steel (see also U.S. Steel) |
| | Gary Works, Gary, IN, 5 mills, 109,000-271,000 t/yr | BW | 1930 to 1938 | ⅜ to 3 | | | | Now a subsidiary of U.S. Steel (see also U.S. Steel) |

## Table C-1. Past Manufacturers of Butt-Welded or Continuous-Welded Pipe

| Manufacturer | Mill/ Capacity | Type Butt or Continuous | Key Dates | Diameter Range, inch | Std API Threaded Pipe | Plain End Line Pipe | Other Information | Historical Information and Predecessor Companies |
|---|---|---|---|---|---|---|---|---|
| | Lorain Works, Lorain, OH, 6 mills | BW | 1920 to 1938 | ⅛ to 3 | | | | 1st built 1894-1895, 1st Bessemer steel, April 1, 1895, 1st O.H. steel Jan. 26, 1909 |
| | National Works, Mckeesport, PA, 6 mills, 192,000-531,600 t/yr | | 1920 to 1945 | ⅛ to 3 | | | 1st steel made Dec 14, 1893 | Now a subsidiary of U S Steel (see also U.S. Steel) |
| | Pittsbrugh works, Pittsburgh, PA, 2 mills, 92,000 t/yr | BW | 1926 (last known year of operation) | ¼ to 2 | | | | Now a subsidiary of U S Steel (see also U S Steel) |
| | Riverside Works, Benwood, WV, 5 mills | BW | 1920 to 1926 | ⅛ to 4 | | | | Sold to Wheeling Steel Corp., Feb. 2, 1928 |
| Reading Iron Company | Reading, PA 21,000 t/yr. | BW | 1920 to 1940 | ⅛ to 2 | | | | |
| Republic Steel Corporation | Youngstown District, Youngstown, OH, 2-4 mills, 152,000-175,000 t/yr | BW | 1920 to 1938 | ⅛ to 4½ | | | | 1920-1938 was known as Republic Iron & Steel Co., Youngstown, OH |
| | Youngstown District, Youngstown, OH, 2 mills, 410,000 t/yr. | CW | 1939 to 1964 | ½ to 4 | | | | |
| Sawhill Tubular Products | Mercer Pipe Div., Sharon, PA, 100,000-120,000 t/yr | CW | 1954 to 1970 | ¼ to 4 | | | Mercer Tube & Mfg. Co. to Jan. 1954 | |
| Sharon Tube Company | Sharon, PA, 5,400 to 48,000 t/yr | CW | 1935 to 1970 | ⅛ to ¾ | | | | |
| Steel Company of Canada, Ltd | Works at Welland, Ontario 300,000 t/yr | CW | Began by Page-Hersey, 1939 | | | | | |
| | McMaster Works, Contrecoeur, Quebec, 1 mill | CW | 1960 to 1979 | ⅛ to 4 | | | Grade A25 | |
| | St. Henry Works, Montreal, 1 mill, 22-51,000 t/yr | Unknown | 1935 to 1957 | ¼ to 4 | | | | |
| | Ontario, Canada, 1 mill, 18,000-25,000 t/yr | BW | 1920-1930 | ¼ to 4 | | | | |
| U.S. Steel Corporation, Pittsburgh, PA | Lorain Works, Lorain, OH, 1 6 mills, 190,000-503,000 t/yr | CW | 1934 to at least 1979 | ⅛ to 4 | | | Grade A25 | |

Table C-1. Past Manufacturers of Butt-Welded or Continuous-Welded Pipe

| Manufacturer | Mill/ Capacity | Type Butt or Continuous | Key Dates | Diameter Range, inch | Std API Threaded Pipe | Plain End Line Pipe | Other Information | Historical Information and Predecessor Companies |
|---|---|---|---|---|---|---|---|---|
| | Fairless Div., Fairless Hills, PA, 1-2 mills, 407,000 t/yr | CW | 1957 to 1970 | ½ to 4 | | | | |
| | Fairless Works, Morrisville, PA, 2 mills, 345,000 t/yr | CW | 1954 (last known year of operation) | ½ to 4 | | | | |
| Wheeling Steel Corp | Wheeling, WV, Benwood Works, 188,000 t/yr | BW | 1928 to 1938 | ½ to 4 | | | | |
| | Wheeling, WV, Benwood Works, 360,000 t/yr | CW | 1939 to 1968 | ½ to 3 | | | | |
| Youngstown Sheet & Tube Co | Campbell Works, Campbell, OH, 5-6 mills, 160,000 t/yr | BW | 1920 to 1938 | ¼ to 3 | | | | Inc Nov 23, 1900 |
| | Campbell Works, Campbell, OH, 1 mill, 180,000 t/yr | CW | 1939 to 1970 | ¼ to 3 | | | | |
| | Indiana Harbor Works, East Chicago, IN, 2 mills, 100,000 t/yr | BW | 1926 to 1938 | ¾ to 4 | | | | |
| | Indiana Harbor Works, East Chicago, IN, 2 mills, 348,000 t/yr | CW | 1938 to 1970 | | | | | |
| | Evanston Works, Evanston, IL, 2 mills, 65,000- 125,000 t/yr | BW | 1926 to 1938 | ¼ to 3 | | | | |
| | Zanesville Works, Zanesville, OH, 2 mills, 120,000 t/yr | BW | 1926 (last known year of operation) | ¼ to 3 | | | | |

Table C-2.  Current Manufacturers of Continuous-Welded Pipe

| Manufacturer | Mill/ Capacity | Key Manufacturing Steps | Key Dates | Hydrotest Info | Diameter Range, inch | Wall Thickness Range, inch | Grades | Lengths | Non Destructive Inspection and Quality Factors |
|---|---|---|---|---|---|---|---|---|---|
| Laclede Steel CO., St Louis, MO | Alton Works | Skelp passing through the preheat and furnace is brought almost to the melting point before entering the hugh speed mill  Manufacture of pipe takes place at speeds up to 800 ft./min | Began 1911, Switched from BW to CW—year unknown | 3000 psi | ½ to 4 | | A25 | 25 | |
| Sawhill Tubular Div, Armco, Sharon, PA | | 4½" mother tube is cont welded and stretch reduced for various sizes | Began 1970 | 700 to 1200 psi | ⅜ to 4 | 0.068 to 0.237 | A25 | 45 ft. max. | |
| Wheatland Tube Co., Collingswood, NJ | 250,000 t/yr | Ladle % Max C, 21 | 1931 began operation, 1963 a new mill was installed | Yes | ⅛ to 4 | 0.068 to 0.237 | A25 | 18 to 46 ft | |

Table C-3. Manufacturers of Lap-Welded Pipe

| Manufacturer | Mill/ Capacity | Key Dates | Diameter Range, inch | Std API Threaded Pipe | Plain End Line Pipe | Other Information | Historical Information and Predecessor Companies |
|---|---|---|---|---|---|---|---|
| Allegheny Steel Co | Brackenridge Plant, Brackenridge, PA, 1-2 furnaces, 18,000 t/yr | 1920 to 1945 | 2 to 6 | | | | 1945 or before became Allegheny Ludlum Steel Corp. In 1948 or before headquarters was in Pittsburgh, PA. |
| Bethlehem Steel Corp. | Sparrows Point, MD 2 furnaces 100,000 t/yr | Began operating in 19... dismantled in 1957 | 2 to 8 / 6 to 16 | | | | 1860 Rider, Pittsburgh 1898 Rider-Conley Manufacturing, Co 1916 Rider/Marshall Merger 1931 Merged into Bethlehem |
| | Coatsville, PA. 4 furnaces boiler tubes | Began operating discontinued by ..... | 1½ to 6 | | | | |
| A.M. Byers Company | Ambridge, PA Closed in 1969 75,000 t/yr | | | | | Sizes larger than 14-inch were cold-formed from plate and hammer-welded or fusion-welded lengths of 16 to 22 ft sizes to 30 inch available | Established as a wrought iron producer in 1864 Red spiral painted on iron pipe to distinguish it from steel pipe. Also rolled, letter "Byers iron" or "Byers Steel". |
| | South Side, Pittsburgh, PA, 2-4 mills, 90,000 t/yr | 1938 to 1964 | 1½ to 14 | Steel and wrought iron standard wall thicknesses and test pressures | Regular weight and extra strong wrought iron only | | |
| Central Tube Co | Pittsburgh, PA, 1 furnace, 64,700 t/yr | 1935 to 1938 | 2 to 13¾ | | | | |
| Jones & Laughlin Steel Corp | Aliquippa Works, Woodland, PA, 2-3 furnaces, 180,000- 380,000 t/yr | ? to 1930 | 1½ to 16 | | | | |
| | Aliquippa Works, Aliquippa, PA, 1-3 furnaces, 75,000- 204,000 t/yr | 1935 to 1957 | 1½ to 16, 1935-1945, 2 to 4, 1948-1957 | | | | |
| National Supply Company | Works at Etna, PA, 1-2 furnaces, 45,000- 137,000 t/yr | 1920 to 1957 | 1½ to 24 | | | | Was Spang, Chalfant and Company, Inc to Oct 23, 1937(Works at Etna was Etna Iron & Tube Works, Etna, PA) |
| National Tube Co | Continental Works, Pittsburgh, PA, 1 furnace, 92,000 t/yr | 1920 (last known year of operation) | 2 to 8 | | | | Subsidiary of U S Steel Corp |
| | Gary Works, Gary, IN, 3 furnaces, 153,800 t/yr | 1930 to 1938 | 3¼ to 22 | | | | |
| | National Works McKeesport, PA, 1-12 furnaces, 104,000- 140,000 t/yr | 1920 to 1938 | ½ to 36, 2 to 24, 3¼ to 24, 3¼ to 16 | | | | 1st steel made Dec 14, 1893 |

Table C-3. Manufacturers of Lap-Welded Pipe

| Manufacturer | Mill/ Capacity | Key Dates | Diameter Range, inch | Std API Threaded Pipe | Plain End Line Pipe | Other Information | Historical Information and Predecessor Companies |
|---|---|---|---|---|---|---|---|
| | Pennsylvania Works, Pittsburgh, PA, 7 furnaces, 290,000 t/yr black pipe | 1920(last known year of operation) | 2 to 16 | | | | |
| | Pittsburgh, Works, Pittsburgh, PA, 1 furance, 92,000 t/yr | 1926 (last known year o | 2 to 8 | | | | |
| | Riverside Works, Benwood, WV, 2 furnaces, 128,000 t/yr | 1920 to 1926 | 2 to 6 | | | | Sold to Wheeling Steel Corp. Feb. 2, 1928 |
| | Syracuse Works, Syracuse, NY, 2 furnaces, 803,000 t/yr black pipe | 1920 to 1926 | 2 to 4 | | | | |
| | Loran Works, Loran, OH, 2 to 7 furnaces, 130,000 t/yr | 1920 to 1928 | 1-1/2 to 20, 2 to 20, 2 to 22, 2 to 9- 1/4 | | | | Subsidiary of U.S. Steel Corp., 1st bult 1894-1895, 1st Bessemer steel Apr. 1, 1895, 1st O.H steel Jan. 26, 1909 |
| Reading Iron Company | Reading, PA, 100,000 t/yr | 1920 to 1940 | 1¼ to 20 | | | | |
| Republic Steel Corporaton | Youngstown Works, Youngstown, OH, 2-6 furnaces, 228,000- 270,000 t/yr? | 1920 to 1938 | 2 to 16 | | | | Inc in 1899 Was Republic Iron & Steel Co, Youngstown, OH |
| South Chester Tube Co | Chester, PA, 108,000 t/yr | 1930 to 1960 | 2 to 13¾ | | | | |
| Wheeling Steel Corp | Wheeling, WV, Steubenvile, OH (together) 136,000 t/yr | 1930 to 1945 | 1½ to 12- 3/4 | | | | |
| Youngstown Sheet & Tube Co | Youngstown, OH, 5 furnaces, 450,000- 600,000 t/yr | 1920 to 1926 | 2 to 20 | | | | Inc on Nov 23, 1900 |
| | Campbell Works, Campbell, OH, 2-4 furnaces, 100,200- 260,000 t/yr | 1930 to 1948 | 2 to 20, 4½ to 20 | | | | |

Table C-3. Manufacturers of Lap-Welded Pipe

| Manufacturer | Mill/ Capacity | Key Dates | Diameter Range, inch | Std API Threaded Pipe | Plain End Line Pipe | Other Information | Historical Information and Predecessor Companies |
|---|---|---|---|---|---|---|---|
| Indiana Harbor Works, East Chicago, IN, 1-3 furnaces, 79,200-220,000 t/yr | | 1926 to 1948 | 2 to 20, 2 to 28, 2¾ to 8⅝ | | | | |
| Zanesville Works, Zanesville, OH, 2 furnaces, 120,000² t/yr | | 1926 (last known year of operation) | 2 to 16 | | | | |

## Table C-4. Seamless Pipe Facilities Past and Present.

| Manufacturer | Mill/Capacity | Key Manufacturing Steps | Key Dates | Diameter Range, inches | Wall Thickness Range, inches | Grades | Non Destructive Inspection and Quality Factors |
|---|---|---|---|---|---|---|---|
| Algoma Steel Corporation, Ltd., Sault Ste. Marie, Ontario, Canada | Mill No. 1 | Traditional plug mill computerized and automated | Operated by Mannesmann 1957 to 1971; CURRENT | 7 to 12.75 | | up to X70 | Raw materials: steel, billets made on-site, BOF Q&T facility adjacent to tube mills. |
| | Mill No. 2 300,000 tons/yr | Retained mandrel process in tandem with--22 stand stretch reduction | Began operating in 1987 CURRENT | 2 to 7 | | | |
| American Seamless Tube Corp | Los Angeles, CA | No Information | API Licensed 1930, 1931. No longer producing line pipe | | | | |
| Armco Steel Corp., Ambridge, PA | Mill No. 1 | Traditional plug mill | 1913 Standard Seamless Tube Company<br>1928 purchased by Spang-Chalfant and Co. No longer producing line pipe. | 4.5 to 14.375 | | | Blooms from Ashland, KY and Butler, PA works. Bar mill at Ambridge made the Billets. |
| | Mill No. 2 | Traditional plug mill | 1930 Purchased by National Supply Co<br>1958 Purchased by Armco Steel Co<br>No longer producing line pipe. | 2.375 to 5.563 | | | |
| Babcock & Wilcox Co. ( tubing operations terminated in 1988 except for reopening of Ambridge by Koppel in 1991) | Beaver Falls, PA | 1 plug mill and Assel mill | Originally Pittsburgh Seamless Tube goes back before 1907 when acquired by B&W-1930- 3 hot mills with a 4th in 1941. No longer producing line pipe. | 3.25 to 9.625<br>1.3125 to 4.333<br>2.375 to 6 | | | Continuous casting started in 1950 |
| | Milwaukee, WI | 1 plug mill; 1 mandrel mill (renovated in 1979) | No longer producing line pipe | 3.25 to 7.625<br>1.3 to 3.25 | | | |
| | Ambridge, PA<br>200,000 tons/yr | Mandrel mill, stretch reduction mill 70 ton/hr rotary furnace<br>Transval mill for mechanical tubing | Installed mid-70's, Purchased by Koppel in 1990<br>CURRENT (Koppel) | 1.9 to 4.5<br>3⅛ to 8½ | 0.150 to 0.650<br>o/t ratios 3.5:1 to 15:1 | | Electric furnace steel, ladle refined and continuously cast |
| Cameron Iron Works | Houston, TX | By extrusion | 1968 to 1987. No longer producing line pipe. | | | | |
| Colorado Fuel & Iron Company | Pueblo, CO | Double-piercing in Mannesmann piercer, 2 plug mills, 2 reeling mills, sizing mill and stretch reducing mill (Aetna Standard mill 200,000 tons/yr) 70 ton/hr rotary furnace. Original pipe mill equipment still in use, but steelmaking has changed. Formerly open hearth or basic oxygen steel, ingot casting, blooms converted to billets after scarfing. Billets peeled and scarfed | Installed 1953 owned by Oregon Steel since 1993<br>CURRENT | 6⅝, 8⅝<br>10.75 plus other sizes for OCTG | | A through X60 | Electric furnace steel, continuous cast since 1976. 6¼ and 7½ inch rounds. ½ of production is heat treated. Amalog MFL inspection --10,000 psi hydrostatic tester |
| Copperweld Steel Co | American Seamless Tubing Facility, Baltimore, MD | 4½" square billets continuous cast, reheated in induction furnace. Formed to round billet, trepanned hole, press formed extruded tube | API License only in 1985. No longer producing line pipe. | | | Main focus was not on line pipe but OCTC | Glass used as lubricant for extrusion |
| | Ohio Steel Tube Facility, Shelby, OH | Rebuilt piercing mill and Assel mill in 1993 | Started as Ohio Seamless Tube Co. early in the century after originally being part of U.S. Steel. Acquired by Copperweld in 1952. No longer producing line pipe. | ½ to 7 | | | |
| Globe Steel Tubes Co. | Milwaukee, WI, 125,000 t/yr. | 2 Mannesmann piercers | 1935 to 1956<br>No longer producing line pipe | ¼ to 7⅝ | | A & B only | |
| Jones & Laughlin Steel Corp., Aliquippa, PA | Mill No. 2 | Rounds rolled on 30-inch round mill, end charged furnaces, 2 Mannesmann rotary piercers, plug mill, 2 reeling mills, sizing mill. | Placed in operation in July 1927<br>No longer producing line pipe. | 5½ to 14 | | | Open hearth steel, cast ingots until 1967, then basic oxygen steel and continuous casting used to form billets |
| | Mill No. 3, 480,000 tons/yr | Mannesmann rotary piercier, plug mill, 2 reeling mills, two hot reduction mills | Placed in operation in July 1928<br>No longer producing line pipe. | 2⅜ to 6 | | | |

**Table C-4. Seamless Pipe Facilities Past and Present.**

| Manufacturer | Mill/Capacity | Key Manufacturing Steps | Key Dates | Diameter Range, inches | Wall Thickness Range, inches | Grades | Non Destructive Inspection and Quality Factors |
|---|---|---|---|---|---|---|---|
| Jones & Laughlin Steel LTV | Indiana Harbor Works | Rotary furnace heating of rounds from Pittsburgh or Aliquippa. 2 Mannesmann rotary piercers, plug mill, 2 reeling mills, sizing mill. Quenching and tempering for certain grades | Originally a Youngstown Sheet and Tube Co. facility. No longer producing line pipe. | 4½ to 9⅝ | 0.237 to 0.337 small OD 0.312 to 0.545 large OD | | Electric furnace steel or basic oxygen steel to make blooms, bloom hot-scarfed, rolled into rounds, smaller rounds machined on peelers at Aliquippa. 10,000 psi hydrostatic tester |
| Koppel Steel | Ambridge, PA | See Babcock and Wilcox | | | | | |
| North Star Steel Co. | Youngstown, OH facility 250,000 t/yr | Piercing mill, Diescher elongator, 6-stand retained mandrel mill | Purchased from Hunt Energy in 1986. Revamped and started 1988. | 4½ to 9⅝ | | | |
| Phoenix Steel Corp. (formerly Barium Steel Corporation, known as Phoenix Iron and Steel Corp. prior to May 1959) | Phoenixville, PA | Rotary furnace, Pilger mill | In operation by 1956 | 7½ to 16 | 0.322 to 2.875 | up to X52 | |
| Pittsburgh Steel Company | Allenport, PA | Billets produced at Monessen, PA plant, peeled at Allenport, heated in two successive furnaces, pierced by Mannesmann rotary piercer, formed in plug mill and one of two reelers. A Pilger mill was also used early on. A Diegcher mill was apparently added latter. | Original mill installed in 1921, replaced in 1944. No longer producing line pipe. | 2⅞ to 7⅝ | | | Open hearth steel, hot-tapped fully deoxidized ingots, blooming, round billets 2 to 6½ inches |
| Quanex Gulf States Tube Div. | Rosenberg, TX | By extrusion | Started 1963 CURRENT | 0.313 to 2.375 | 0.035 to 0.281 | A, B | |
| Republic Steel Company | Chicago District | One rotary and one roll-down furnace 2 Mannesmann rotary piercers, 1 plug mill, 2 reeling mills, 1 7-stand sizing mill, normalizing and tempering facilities | Started in 1953. No Longer Producing Line Pipe. | 4½ to 9⅝ | | | Inspection equipment installed in early 1980s. Eddy-current to inspect tube rounds before heating and piercing. Sonoscope (MFL) for transverse flaws at OD and ID (i.e. pits, slivers, slugs) Amalog (MFL) for seams and laps on OD surface. |
| Tubos de Aceros De Mexico | Veracruz, Mexico | Sponge iron from direct reduction melted with scrap in electric furnace to make round ingots. Ingots heated in rotary furnace. Pipe made by Mannesmann type piercer, shaped by an elongator and then a Pilger mill. Next step was reheating and either sizing or stretch reduction followed by straightening | Started 1956 CURRENT | 6⅝ to 18, later 1¼ to 18 | 0.140 to 1.250 | A through X65 | |
| United States Steel (includes National Tube Company) | Ellwood, PA Works 250,000 tons/yr | Originally a Stiefel piercer and Pilger mill, changed to 2 Mannesmann piercers, plug mill, 2 reeling mills, and a sizing mill in 1906. By 1970 a Diescher mill had been added to make sizes 1½ to 4 inches | Started in 1895 new mill built 1933 (not API certified by 1979 and believed to be closed down) | 1½ to 4 and 2⅜ to 10.75 | | | |
| | Fairfield, AL, Works 600,000 tons/yr | Press piercing mill. Square bloom is heated to 23,000 F, descaled, pierced and rounded to a heavy walled tube. Next the tube is rotary elongated over a plug and bar followed by rolling in a multistand mandrel mill. It is then reheated and stretch-reduced. Heat treating is done if necessary followed by hot sizing and hot straightening. | Completed, 1983 CURRENT | 4½ to 9⅝ | | | Electromagnetic inspection for longitudinal and transverse flows, VT inspection for wall thickness, eddy current for grade verification. |
| | Gary, IN Works | The nature of the mill process before 1935 has not been determined. From 1935 to 1942. Apparently 2 Mannesmann mills (piercers, lug mill, rollers, sizers) existed. One, apparently the larger of the two, was then dismantled. From 1942 until at least 1979, at least one and as many as three mills Mannesmann, mandrel and Assel were operated. The Gary Works apparently still produces seamless pipe but not API line pipe | Built in 1916 by Gary Tube Co., capacity upgraded 1928-1930. No longer producing line pipe | 4½ to 13 to 1935<br>3½ to 24 1935 to 1942<br>1 to 8⅝ 1942 through 1979 | | | |

**Table C-4. Seamless Pipe Facilities Past and Present.**

| Manufacturer | Mill/Capacity | Key Manufacturing Steps | Key Dates | Diameter Range, Inches | Wall Thickness Range, Inches | Grades | Non Destructive Inspection and Quality Factors |
|---|---|---|---|---|---|---|---|
| | Lorain, OH Works | 2 Mannesmann piercers, plug mill, 2 reeling mills, sizing mill. Sizes 16-inches and up subjected to reheating and rotary expansion after plug mill and prior to reeling. Cold expansion by internal traveling plug is available. | Seamless mill constructed 1928-1930 CURRENT | 1930 to 1935 2⅜ to 13¾; 1935 to 1945 2⅜ to 24, 1945 to present 2⅜ to 26 | | | Continuous cast billets rotary furnace electromagnetic inspection |
| | National Works, Mckeesport, PA | 2 Mannesmann piercers, plug mill, 2 reeling mills, sizing mill Sizes 16-inches and up subjected to reheating and rotary expansion after plug mill and prior to reeling. | Seamless mill constructed 1933 No longer producing line pipe. | 3½ to 24 | | | |
| Youngstown Steel & Tube Co | Campbell Works, Youngstown, OH 300,000 tons/yr | 1 Pilger mill with mannesmann piercer 1 Traditional plug mill with stretch reduction for smaller sizes. | Began in 1926 Phased out Pilger mill No longer producing line pipe | 2⅜ to 14 | | | |
| | Indiana Harbor | See J&L/LTV | | | | | |

## Table C-5. Past Manufacturers of ERW Pipe.

| Manufacturer | Mill/ Capacity | Type | Key Manufacturing Steps | Key Dates | Cold Expanded Yes/No | Hydrotest Info | Diameter Range, inch | Wall Thickness Range, inch | Grades | Lengths | Non Destructive Inspection and Quality Factors |
|---|---|---|---|---|---|---|---|---|---|---|---|
| Aceros Alfa | Monterrey, Mexico | Unknown | Unknown | 1962 to 1973 | | | 6⅝ | | | | |
| Armco Steel Corp., Ambridge Works | 200,000 t/yr | LF "Spangweld" | Coils were slit, edges shot blasted and trimmed forming was by cage rolls, fin pass, pipe was welded, reheated, stretch-reduced, and cut to length | Approximately 1964 to 1967 | No | Not available | Welded in sizes 5-3/4, 7¼, and 8⅝, stretch reduced to ½ to 4" | Not Available | Not Available | Not Available | |
| Beall Pipe & Tank Co. | Portland, OR | Unknown | Unknown | 1956 to 1982 | No | 3 to 16 | up to 0.250 | up to X92 | | | |
| Bethlehem Steel Co., Sparrows Point, MD | Small OD up to 225 fpm | HFC 2800 kw 450 kHz | Coils were slit, side trimmed, formed nearly round by 4 preforming rolls, held by cage rolls, prepared for welding by 3 fin-pass stands, welded, flash trimmed, seam annealed by two 3000-cycle induction heaters inspected, sized, and cut to length hydrotested, straightened, beveled | Began operation in 1963, discontinued in 1982 | No | API Minimum | 2.375 to 6.625 | 0.065 to 0.280 | A through X52 | up to 50 ft for sizes to 4½-in., up to 65 ft for 6-in. | Seam inspected by eddy current immediately after annealing, full-body eddy current inspection after straightening but before hydrostatic test. |
| | Large OD up to 120 fpm | LF Yoder Mill | Coils were leveled, edge trimmed by rotary shear, edges shot-blasted, formed by 4 convex-concave rolls, 2 vertical rolls, 3 fin-pass rolls, welded, flash-trimmed, seam-inspected, seam "post-tempered" to 1200 F by Tocco induction unit, heater, cut to length, sized, straightened, beveled, flushed with water, hydrotested. | Started in 1957, discontinued in 1970 | No | API Minimum | 5.625 to 16 | 0.188 to 0.500 | A through X52 | up to 65 ft | Seam inspected by ultrasonics after post tempering, additional inspection by means of ultrasonics, magnetic flux leakage, and/or eddy current was apparently available by agreement. |
| | Large OD | HFC 560 kw | Coils were leveled, edge trimmed by rotary shear, formed by a forming stand, cage rolls and 4 fin-pass rolls, welded, flash-trimmed, seam-annealed by 2-350 kw and 2-300 kw induction units, sized, cut to length, inspected, hydrotested, inspected again. | Began operation in 1970, discontinued in 1983 | No | API Minimum | 5.5625 to 16 | 0.188 to 0.500 | A though X60 | up to 65 ft | Seam inspection by ultrasonics after seam-annealing, prior to and after hydrostatic test. |
| Big Inch Pipe Corporation, Calgary, Alberta, Canada | 25 to 75 fpm | HFC 280 kw 400 kw Thermatool by Cal Metal Corp | 60 ft plates trimmed by rotary shear, formed by a series of convex-concave rolls, welded, flash-trimmed, seam-normalized to 1800 F by low-frequency induction unit, air/water cooled, sized, seam inspected by magnetic flux leakage, mechanically expanded, seam inspected by ultrasonics, beveled, hydrotested | Began operation in 1960, acquired by Canadian (Alberta) Phoenix in 1964 | Yes | 90% of SMYS | 18 to 36 | 0.188 to 0.625 | at least X52 | up to 65 off | Magnetic flux inspection of seam before expansion ultrasonic inspection of seam after expansion |
| Brooks Tube, Ltd. | Brooks, Alberta | | No Information | 1975 to 1983, Acquired by Ipsco in 1983 | No | | 3/4 to 2 | 0.083 to 0.154 | A through X42 | | |
| Bull Moose Tube Co | Gerald, MO | HFC Abby-Etna Mill | | Started in 1975. API License 1980, 1982 | | | 11 Max | 0.375 Max. | | | |
| Cal-Metal Corporation (For other information see Republic and Stupp) | Torrance, CA | HFC 450 kHz | Coils continuous roll-formed, welded, seam-annealed (1400-1500F) by Tocco high-frequency induction units | Began operating 1963, discontinued by 1983 | No | API Minimum | 3½ to 16, later 2⅜ to 18 | 0.109 - 0.432 | A through X60 | | Ultrasonic inspection mentioned in literature |
| Canadian Phoenix Steel & Pipe, Ltd | Edmonton, Alberta, 150,000 t/yr. | Unknown | Not Available | Began in 1960 as Alberta Phoenix. Purchased by Ipsco, Inc., 1973 | No | API Minimum | 3½ to 16 | Unknown | Unknown | Unknown | |
| | Calagry, Alberta | HFC 280 kw, 400 kw thermatool by Cal-Metal Corp | See Big Inch Three Pipe | Began in 1960, as Big Inch Pipe Corp., discontinued some time after 1970 | Yes | API Minimum | 18 to 36 | 0.188 to 0.625 | At least X52 | Up to 65 ft. | |
| | Port Moody, B.C., 130,000 t/yr. | Unknown | Unknown | Began in 1960 as Canadian Western Pipe Mills, Ltd. Acquired byCanadian Phoenix between 1967 and 1970 | No | API Minimum | ½ to 4½ | Unknowm | Unknown | Unknown | |
| Central Steel Tube Co | Clinton, IA, 30,000 t/yr. | | No Information | | | | 4½ to 8½ | 0.250 Max. | | | |
| Consolidated Western Steel Division of U.S. Steel | Provo, UT | LF | Slit coils leveled, skelp formed, welded, flash-trimmed, cooled, straightened, cut to length hydrotested, inspected | Began operating at Provo in 1956 but formerly was operated in Berkeley, CA | No | | 4 to 12.75 | up to 0.188 | | 50 ft. | Skelp rolled at Columbia-Geneva Steel next door |

Table C-5. Past Manufacturers of ERW Pipe.

| Manufacturer | Mill/ Capacity | Type | Key Manufacturing Steps | Key Dates | Cold Expanded Yes/No | Hydrotest Info | Diameter Range, inch | Wall Thickness Range, inch | Grades | Lengths | Non Destructive Inspection and Quality Factors |
|---|---|---|---|---|---|---|---|---|---|---|---|
| Copperweld Steel Co | Ohio Steel Tube Facility, Shelby, OH | LF 1937 to 1986, HFC 1986 | Unknown | API Licensed 1954-1970, 1985-1987 |  |  | ½ to 7½ later ½ to 12-3/4 | 0.029 to 0.625 | A, B |  |  |
|  | Regal Tube Facility, Chicago, IL | Unknown | Unknown | API Licesed 1985-1987 |  |  | 1 to 6 |  | A, B |  |  |
| Fort Collins Pipe Co | Fort Collins, CO |  | No Information | Began in 1963 as Timberline Tube Inc Became Southwestern Pipe of Colorado. Apparently operated under Lone Star's ownership and license between 1975 and 1985 Reappeared in 1985 as Fort Collins Pipe Co. Not on current list. | No | API Minimum | 2¾ to 8⅝ | 0.109 to 0.322 | A, B apparently made X grades as well up to X52 |  |  |
| Fox Steel Pipe | Jacksonville, FL | HFC | Unknown | Began 1962 Acquired by American Steel Pipe in 1963 and moved to Birmingham, AL. |  |  |  |  |  |  |  |
| Geneva Tube | Geneva, NE |  | No Information | API Licensed 1984-1987 |  |  |  |  |  |  |  |
| Jones & Laughlin Steel Corp., Aliquippa, PA | LF McKay Machine Co., 120 cycles, 1100 kw | LF | Coils were leveled, edges trimmed and steel-grit blasted, formed in a 10 stand series of rolls (last 3 probably fin-pass rolls) welded, flash-trimmed, stress-relieved by a Tocco high-frequency induction unit, sized, cooled, cut-off, straightened | Began operating 1957, discontinued by 1965 | No | API Minimum, 3000 psig max. | 6.625 to 12.75 | 0.188 to 0.500 | A through X52 | up to 65 ft | Visual inspection and magnetic particle inspection of ends only. |
|  |  | HFC | Coils leveled, edges trimmed by rotary shear, 10-stand forming with fin-pass roll as last stage, welded, flash-trimmed, ultrasonically inspected, seam-normalized, sized, cut off, straightened, beveled, hydrotested | Began operating 1965, discontinued by about 1980 | No | API Minimum | 4.5 to 12.75 |  | A through X60 |  | Ultrasonic inspection of seam after welding and again after hydrotesting |
| Kaiser | Fontana, CA, 292,000 tons/yr | LF |  |  |  |  |  |  |  |  |  |
|  | 150 fpm 560 kw Torrance Hill | HFC | Coils or cut lengths used to for tubes | Started in 1965 sold to California Steel Industries in 1984 |  | 0.104 to | 4.5 to 16 | 0.104 to 0.508 |  | up to 80 ft |  |
|  | Napa, CA |  |  | Acquired from Basalt Rock between 1954 and 1957 |  |  | 6.625 to 20 |  |  |  |  |
| Lone Star | Mill No 1 40 to 120 fpm, 140 pipes/hr | LF Yoder Mill 4,000 kw | For w t < 0.375 coils were used, for thicker pipe, individual cut lengths were used. Skelp was leveled, edges were trimmed and shot-blasted, skelp was formed into tube in 10 forming stands, welded, flash was trimmed, 3 pull out stands provided straight-ening, lengths were cut by flying cut off, cut lengths were given additional sizing and straightening. Pipe was then full-body normalized to 1650 F, cooled, sized by 2% cold reducing, straightened, inspected, hydrotested, inspected again, marked, and weighed. | Began operation in 1953 | No | API Minimum | 6.625 to 16 | 0.250 to 0.500 | A through X52 | up to 60 ft |  |

## Table C-5. Past Manufacturers of ERW Pipe.

| Manufacturer | Mill/ Capacity | Type | Key Manufacturing Steps | Key Dates | Cold Expanded Yes/No | Hydrotest Info | Diameter Range, inch | Wall Thickness Range, inch | Grades | Lengths | Non Destructive Inspection and Quality Factors |
|---|---|---|---|---|---|---|---|---|---|---|---|
| | Mill No 2 50 to 150 fpm 180 pipes/hr both mills together 350,000 tons/yr | LF | Slit coils were used in a process identical to that described for Mill No 1 above | | No | API Minimum 7000 psi max | 1.9 to 6.625 | 0.125 to 0.344 | A through X52 | up to 60 ft | Full-width coil slit into 3 coils |
| Mobile Pipe Corporation, San Francisco, CA | 8000 ft/day | HFI 450 kHz kw Tocco Div of Ohio Crankshaft Co | Coils leveled, edges scarfed, formed by 3 breakdown passes and 2 fin passes, welded, flash trimmed, air cooled, sized, bent if necessary, fed onto ground as unit moved forward, lengths up to 2000 ft produced continuously | Began operating 1962, received API Cert 1970 but apparently was discontinued by 1975 | No | Field hydrotested after burial | 6.625 to 10.75 | 0.156 to 0.250 | X42 to X52 | | Designed to make ERW pipe in the field from coiled skelp. No seam annealing was provided. Ultrasonic inspection of the seam was performed after sizing |
| National Pipe & Tube Co., Liberty, TX | | HFC Thermatool 600 kw Abbey Etna forming mill | Coils leveled, slit, edge-trimmed, formed, welded, flash-trimmed, full-body normalized and, for some sizes, stretch reduced, sized, straightened, cut off, hydrotested | Began operating in 1976, closed by 1990 | No | API Minimum or more in some cases | 2.375 to 10.75 | 0.109 to 0.375 | A-25 through X56 | up to 60 ft | Ultrasonic and magnetic flux leakage inspection of weld seam after welding, full-body eddy current inspection prior to hydrotest, magnetic flux leakage inspection of weld seam after hydrotest |
| Newport Steel Corporation, Newport, KY | | LF | Slit coils were leveled, shot blasted, edge-trimmed by skiving, formed by a 9-stand series of convex-concave rolls (probably ending with fin-pass), welded, flash-trimmed, seam annealed, by induction unit, sized, cut to length, beveled, hydrotested | Began operations 1951, 1957-1968 Acme-Newport, 1965-1978 Interlake, discontinued by 1980 | No | API Minimum | 4.5 to 8.625 | 0.125 to 0.280 | A through X52 | 20 to 52 ft | NDE not mentioned in available literature, steelmaking by electric furnace |
| Page-Hersey Tubes, Limited, Welland, Ontario, Canada | 15 to 60 fpm 30 pieces/hr | LF 62.5 Hz 1667 kva | Plates leveled, rotary edge trimmed, shot-blasted, formed in stages, welded, flash-trimmed, sized, straightened, inspected, washed, cut off, beveled, cold-expanded, hydrotested. Forming involved 4 breakdown rolls to U-shape, idler passes and 2 fin passes prior to welding | Began operating 1950 Replaced by HFC unit in | Yes, hydraulic | API Minimum | 4½ to 16 | 0.188 to 0.500 | | 17 to 51 ft | |
| Productos Tubulares, S.A., Monclova, Coah., Frontera, Coah., Mexico | Unknown | Unknown | Unknown | In operation around 1970 | | | ½ to 4½, later 5½ to 16 | | up to X60 | | |
| Republic Steel Company (known as Republic Iron and Steel Company until 1938, became LTV Corp in 1978, Youngstown Works) | Mill No 1 60 Hz 6875 kva | LF | Precut lengths of skelp edge trimmed, shot-blasted, formed to a can, welded, sized, straightened | Began operating 1929 | Apparently hydraulic expansion was done for a period of time | API Minimum | 2.375 to 6.625 | | A,B initially, later to X52 | 33 to 52 ft | |
| | Mill No 2 | LF | Precut lengths of skelp edge trimmed, shot-blasted, formed to a can, welded, sized, straightened | Began operating in 1930 Converted to HFC unit in 1961 | Apparently hydraulic expansion was done for a period of time | API Minimum | 4.5 to 8.625 | | A,B initially, later to X52 | 33 to 52 ft | |
| | Mill No 3 | LF | Precut lengths of skelp edge trimmed, shot-blasted, formed to a can, welded, sized, straightened | Began operating in 1930 Converted to HFC unit in 1961 | Apparently hydraulic expansion was done for a period of time | API Minimum | 8.625 to 16 | to 0.500 | A,B initially, later to X52 | 33 to 52 ft | |

Table C-5. Past Manufacturers of ERW Pipe.

| Manufacturer | Mill/ Capacity | Type | Key Manufacturing Steps | Key Dates | Cold Expanded Yes/No | Hydrotest Info | Diameter Range, inch | Wall Thickness Range, inch | Grades | Lengths | Non Destructive Inspection and Quality Factors |
|---|---|---|---|---|---|---|---|---|---|---|---|
| | Mill No 5 300,000 tons/yr | HFC 560 kva Thermatool 450 kHz 2-350 kw seam annealers 960 Hz | Coil was leveled, rotary edge-trimmed, formed by a series of breakdown rolls, cage rolls, and fin pass rolls, welded, flash-trimmed, seam annealed, sized, 1700 to 2000F, air/water cooled, cut to length, straightened, unspected, inspected again | Began operating in 1963, replaced Mill No. 2 and Mill No 3 With some modifications became the current LTV mill | No | API Minimum 5 seconds, 2 cycles of test pressure applied for higher grades | 6.625 to 16 | 0.109 to 0.500 | A through X52 | 24 to 80 ft | Seam inspected after welding by ultrasonic and electro magnetic units, and after hydrotesting by an ultrasonic unit. |
| Southern Pipe & Casing Co | 80,000 t/yr at peak in 1954 | | Unknown | Began 1945 Div of U S Industries Inc 1960-1962 div. of American Pipe & Construction Co, 1965-1969 | Unknown | Unknown | 4 to 14 | Unknown | A, B, and X Grades | Unknown | |
| Southwestern Pipe Inc., Houston, TX | 30,000 t/yr | | Unknown | 1957 to 1973 | Unknown | Unknown | ½ to 4½ | Unknown | A, B and X Grades | Unknown | |
| Standard Tube Co., Detroit, MI | | | Unknown | 1958 to 1985 | | | 6 625 to Max | | | | |
| Standard Tube of Canada, Ltd. Woodstock, Ontario | | Unknown | Unknown | In operation around 1970 to 1975 | | | 4 mill to 3½ 1 mill to 5 | | A,B | | |
| Valmont Industries, Valley, NE | | | No Information | | | | 6¾ to 10-3/4 | | A, B, and X Grades | | |
| Youngstown Steel & Tube, Youngstown, OH, Brier Hill Works | Original large OD Mill | d c welder | Plates formed in stages to tubes, welded, exterior flash-trimmed, cooled, sized straightened, cropped, beveled, hydrotested, descaled, inspected | Began operating in 1930 replaced by 1948 | No | API Minimum | 16 to 26 | 0.250 to 0.50 | B | | Open hearth steel, early pipe not seam-annealed or ID flash-trimmed |
| | Original small OD mill | d c welder | Believed to be similar to the large OD mill | In operation by 1935, replaced by 1948 | No | API Minimum | 6 625 to 12 75 | | B | | |
| | Final mill 336,000 tons/yr | d c welder | 53 ft pieces of skelp trimmed to width, edges slightly beveled, shot-blasted, progressively formed to cylinder, welded, flash-trimmed OD and ID, inspected, seam heat-treated, cold expanded (larger sizes only), hydrotested | In operation by 1948, closed down by 1980 | Yes, for larger sizes 1¼% | API Minimum, 3000 psi max | 6 625 to 22 | 0.125 to 0.812 | A through X60 | up to 52 ft | Seam magnetic flux inspected after welding and again after hydrotest |

## Table C-6. Current Manufacturers of ERW Pipe

| Manufacturer | Mill/ Capacity | Type | Key Manufacturing Steps | Key Dates | Cold Expanded Yes/No | Hydrotest Info | Diameter Range, inch | Wall Thickness Range, inch | Grades | Lengths | Non Destructive Inspection and Quality Factors |
|---|---|---|---|---|---|---|---|---|---|---|---|
| American Steel Pipe, Birmingham, AL | Mill No. 1 | HFC(a) 600 kw Thermatool | Coils trimmed by rotary shear, preforming rolls, jacks and cage form rolls, welding, trimming, 4-500 kw Tocco seam annealers, low temperature stress relief | Operation started 1963 | No | 100% of SMYS for 20 sec | 10-3\4 to 20 | 0.188 to 0.500 | B through X70 | up to 100 ft | 100% seam inspection by ultrasonics, full-body ultrasonic inspection available, dry magnetic particle inspection |
| | Mill No. 2 | HFC 600 kw Thermatool | Coils trimmed by rotary shear, preforming rolls, jacks and cage form rolls, welding, trimming, 4-500 kw Tocco seam annealers, low temperature stress relief | Operation started 1989 | No | 100% of SMYS for 20 sec. | 16 to 24 | 0.250 to 0.750 | B through X70 | up to 100 ft | 100% seam inspection by ultrasonics, full-body ultrasonic inspection available, dry magnetic particle inspection |
| Bellville Tube Corporation, Bellville, TX | | HFC(b) | Coils are slit on site, forming by break down rolls and fin pass rolls welding, trimming, air/water cooling, sizing, cutting to length, full-body normalizing, air cooling, straightening | Operation started 1980, (part of Quanex) | No | Meets API Minimum | 2⅜ to 4½ | 0.154 to 0.337 | B, X42 X52 | 35 to 45 ft | 100% ultrasonic inspection of seam prior to full-body normalizing. Full-body inspection by eddy-current and 2nd 100% ultrasonic inspection of seam prior to full-body normalizing. Raw material limited to fully-killed continuous cast steel with inclusion shape control. |
| California Steel Industries, Inc., Fontana, CA | 150,000 tons/yr | HFC 280 kw 400 hz | Coils from concast slabs, preforming rolls with edge bending, cage forming, welding, trimming, 1750 F seam heat treating | Operation started 1988, by renovating a former Kaiser facility | No | Meets API Minimum, 100% SMYS for 10 sec. 3,000 psi max. | 4½ to 16 | 0.156 to 0.406 | B through X65 | up to 65 ft | 100% ultrasonic inspection of seam, full-body inspection by magnetic flux leakage. Raw material restricted to basic oxygen or electric furnace steel, fully-killed, fine grained, continuously cast with a limit of 0.36% on the sum of Cu+Ni+Cr+Mo. |
| Camp Hill Corporation, McKeesport, PA(U.S. Steel) | 70 - 110 fps | HFC Thermatool 450 kHz | Coils trimmed by rotary shear and skiver, forming by 4 breakdown rolls, 3 positioner rolls, 3 fin pass rolls, welding, trimming, seam normalizing 1750 F, air/water cooling, sizing, flying cutoff, straightening after hydrostatic test. | Camp Hill Corp. operates the mill started by U.S. Steel in 1964. Camp Hill was formed in 1988 | No but U.S Steel did at one time | Meets API minimum-- 5000 psi max. | 8⅝ to 20 | 0.172 to 0.375 | B through X70 | up to 65 ft | 100% ultrasonic inspection of seam prior to heat treatment and again after hydrostatic testing, full-body magnetic flux leakage inspection. Raw material apparently unrestricted |
| Camrose Pipe Company, Camrose, Alberta, Canada | | HFC Thermatool 560 kw | Coils are leveled and edge trimmed. Preforming is done in 4 convex-concave rolls followed by a series of 3 vertical rolls and 3 fin pass rolls. Welding, flash trimming, seam annealing, sizing, rotary cut off | Started in 1963 by Stelco/Page-Hersey 60% sold to Oregon Steel Mills in 1992 | Yes (Hydraulic for 8-in. and up) | API Minimum | 4½ to 16 | 0.125 to 0.500 | A through X65 | | 100% ultrasonic inspection of seam after annealing and again after hydrostatic test. All skelp from Stelco or Oregon Steel |
| Geneva Steel, Vineyard, UT | 120,000 tons/yr | HFC | | Purchased from U.S. Steel in 1987 | | | 6⅝ to 16 | | | up to 80 ft | |
| Hysla SA de CV San Nicolas de los Garza, Mexico | | HFC | Coils, forming, welding, full-body normalizing, stretch reduction, cooling, cutting, straightening | | | 3000 psi max | 2⅜ to 4½ | 0.154 to 0.237 | A-25 through X52 | | |
| Ipsco, Inc. | Calgary | HF 600VT Thermatool | Preslit coil processed through preform rolls, cage forming, 2 fin stands, flash trim, 1200 KW seam annealers, air cooled section sizing and rotary disk cutoff | Commissioned 1982 | | 100% of SMYS 10,000 psi | 4½ to 10-3/4 | 0.125 to 0.500" | | | 100% ultrasonic seam inspectionon mill line and psot hydro |
| | Edmonton | HF 280 VT Thermatool | Inline slitter, vertical accumulator, preform rolls, cage forming, 3 fin stands, flash trim, 500 KW, seam anealers, sizing and parting tool type cutoff | Purchased from Phoenix Steel | No | 3,000 psi capacity | 4½ to 16 | 0.125 to 0.375 | | | 1005 ultrasonic seam inspection |
| | Red Deer | HF 300 VT Thermatool | Preslit coil accumulator, preform rolls, cage forming, 3 fin stands, flash trim, 600 KW seam annealler, sizing section and fricition saw cutoff for HSS in site range | Purchased from RAM Steel | | 4,500 psi capacity | 2⅞ to 12-3\4 | 0.105 to 0.500 | | | 100% ultrasonic seam inspection plus in line full body EMI surface inspection in pipe form |
| | Regina 2" mill | 160 VT Thermatool | Preslit coil, preform rolls, fin stands, 300 KW seam annealler, friction cutoff HSS in size range | Started 1956 as Prairie Pipe Manufacturing Co. Upgraded in 1992 | | 3,000 psi capacity | 1½ to 2⅜ | 0.083 to 0.302 | | | 100% ultrasonic seam inspection |
| | Regina 24" mill | 600 VT Thermatool Welder | Self threading mill, in line edge slitting, preform stands, edge preform rolls, cage forming, flash triming, 100% mill ultrasonics, 1,000 KW seam annealler, 4 stand sizing, rotary parting tool cutoff | Built 1994 | | 3,000 psi capacity | 14 to 24 | 0.250 to 0.500 | | | 100% ultrasonic seam inspection, full body EMI |
| | Camanche | HV 300 VT Thermatool | Preslit coil, horizontal accumulator, preform rolls, cage forming, 3 fin stands, flash trim, seam annealler, sizing, rotary disk cutoff | | | 5,000 psi capacity | 2⅜ to 8⅝ | 0.125 to 0.400 | | | 100% ultrasonic seam inspection |

**Table C-6. Current Manufacturers of ERW Pipe**

| Manufacturer | Mill/ Capacity | Type | Key Manufacturing Steps | Key Dates | Cold Expanded Yes/No | Hydrotest Info | Diameter Range, inch | Wall Thickness Range, inch | Grades | Lengths | Non Destructive Inspection and Quality Factors |
|---|---|---|---|---|---|---|---|---|---|---|---|
| Lone Star Steel, Lone Star, TX | Mill No 1 | HFC | Coils leveled, edge-trimmed, 3 preform rolls, cage forming rolls, 4 fin-pass rolls, welding, flash trimming, rotary cut off, full-body normalizing by induction to 1650 F, hot reduction, air cooling, sizing, straightening. | Operation started in 1968 | No | 90% SMYS for 10 sec | 8⅝ to 16 | 0.188 to 0.625 | B through X70 | | 100% ultrasonic seam inspection after welding and is available after hydrostatic testing as is full body ultrasonic and/or magnetic for leakage inspection. 100% full-body magnetic flux/leakage inspection before hydrotesting, electric-arc furnace steel, bottom-teemed ingots, rolled slabs although continuous cast slabs from other suppliers may be used |
| | Mill No. 2 | HFC | Coils leveled, edge trimmed, 3 preform rolls, cage forming rolls, and 4 fin-pass rolls, welding, flash trimming, rotary cut off, full-body normalizing by induction to 1650 F, stretch reduction, air cooling, sizing straightening | | No | 90% SMYS for 10 sec | 2⅜ to 6⅞ | 0.154 to 0.500 | B through X70 | 48 | 100% ultrasonic seam inspection after welding and is available after hydrostatic testing as is full-body ultrasonic and/or eddy current inspection. 100% full body eddy current inspection before hydrotesting. Electric-arc furnace steel, bottom-teemed ingots, rolled slabs although continuous cast slabs from other suppliers may be used. |
| | Seam-Annealed ERW | HFC | Slit coils are leveled, (formed by 4 convex-concave rolls, 3 fin-pass rolls, welded, flash-trimmed, induction seam-annealed above 1250 F air-cooled, sized, straightened, cut to length | Operation started 1992 | No | API Minimun | 4½ to 8⅝ | 0.156 to 0.322 | B, X42 | 60 | 100% ultrasonic seam inspection after welding |
| LTV Steel Tubular Products Company | Cleveland, OH | HFC (4 welders) | Coils leveled, rotary sheared, formed, welded, flash-trimmed, sized, cut to length. Forming appears to involve 3 preform rolls, 1 pair of vertical rolls, and 3 fin-passes. | | No | API Minimum | 3½ to 8⅝ | 0.156 to 0.281 | B, X42 | | 100% ultrasonic seam inspection after welding |
| | Counce, TN | HFC | Coils leveled, rotary sheared, formed, welded, flash-trimmed, seam-annealed, sized, cut to length. Forming appears to involve 3 preform rolls, 1 pair of vertical rolls and 3 pin-pass rolls. | Former Cal-Metal Mill acquired in the late 1960s | No | API Minimum | 2⅜ to 4½ | 0.125 to 0.300 | B through X65 | 17 to 50 ft | 100% ultrasonic seam inspection after welding. Basic oxygen steel making, vacuum degassing, continuous cast slabs although skelp may be used from other suppliers. |
| | Youngstown, OH | HFC 560 kw 180 pipes/hr 380,000 tons/yr | Coils leveled, rotary sheared, formed, welded, flash-trimmed, seam-annealed, sized, cut to length. Forming appears to involve 3 preform rolls, 1 pair of vertical rolls and 3 pin-pass rolls | Began operating in 1963 | No | API Minimum | 8⅝ to 16 | 0.188 to 0.500 | B through X65 | 17 to 60 ft | 100% ultrasonic seam inspection after welding. Basic oxygen steel making, vacuum degassing, continuous cast slabs although skelp may be used from other suppliers. |
| Maverick Tube Corporation, Conroe, TX and Hickman, AR | | | | | | | 2.375 to 8.625 | 0.154 to 0.500 | up to X70 | 44 ft | |
| Newport Steel Corporation, Newport, KY | 8-inch Wean Mill (total capacity) | HFC Thermatool 280 kw 3 Tocco seam annealers | Slit coils are leveled, roll-formed, welded, flash-trimmed, seam annealed, sized, cut to length by flying cut off, straightened, beveled | Formerly Interlake, Acme-Newport, began operating in 1983 | No | API Minimum | 4½ to 8⅝ | 0.156 to 0.322 | B through X52 | | 100% ultrasonic seam inspection and 100% full-body magnetic flux leakage inspection after hydrostatic testing. Electric furnace steel, continuous-cast slabs |
| | 16-inch Wean mill, 400,000 tons/yr (both mills) | HFC Thermatool 600 kw 3 Tocco seam annealers | Coils are leveled, in-line edge-trimmed, welded, flash trimmed, seam-annealed, cut to length, beveled. | Began operating in 1983 | No | API Minimum | 8⅝ to 16 | 0.156 to 0.500 | B through X52 | | 100% ultrasonic seam inspection and 100% full-body magnetic flux leakage inspection after hydrostatic testing. Electric furnace steel, continuous-cast slabs. |
| Northwest Pipe & Casing of Kansas, Atchison, KS | | HFC | | | No | 75% SMYS | 3 through 16 | 0.075 through 0.250 | B, X42 | | |
| Pittsburgh Tube Company | Darlinton, PA | HFI/SRM(c) | | | | | ⅝ to 5 | 0.035 to 0.180 | A,B | | 100% ultrasonic seam inspection |
| Procarsa SA de CV Cuidad Frontera, Mexico | 3 ERW mills | | | | | | 2.375 to 4.5<br>5.5675 to 16 | 0.125 to 0.237<br>0.188 to 0.500 | B<br>B through X60 | | Single and double random lengths |
| Prudential Steel Ltd., Calgary, Alberta, Canada | Mill No. 1 60,000 tons/yr | | | Began in 1966, sold to Dofasco Inc. in 1973, went public in 1994 | | | 2⅜ to 4½ | | | | |
| | Mill No. 2 200,000 tons/yr | HFI | Coil is leveled, formed, welded, flash-trimmed, seam heat-treated, air/water cooled, sized, cut to length, straightened. | Built in 1975 | No | API Minimum | 5⅝ to 12-3/4 | 0.125 to 0.500 | B through X60 | | 100% full-body electromagnetic inspection and 100% ultrasonic seam inspection after hydrostatic test. Skelp supplied by various suppliers |

## Table C-6. Current Manufacturers of ERW Pipe

| Manufacturer | Mill/ Capacity | Type | Key Manufacturing Steps | Key Dates | Cold Expanded Yes/No | Hydrotest Info | Diameter Range, inch | Wall Thickness Range, inch | Grades | Lengths | Non Destructive Inspection and Quality Factors |
|---|---|---|---|---|---|---|---|---|---|---|---|
| | Mill No 3 | HFI | Coil is leveled, formed, welded, flash-trimmed, seam heat-treated, air/water cooled, sized, cut to length, straightened | Built in 1994 | No | API Minimum | 2⅜ to 4½ | 0.125 to 0.438 | B through X60 | | 100% full-body electromagnetic inspection and 100% ultrasonic seam inspection after hydrostatic test. Skelp supplied by various suppliers |
| Sawhill Tubular Div., Armco, Sharon, PA | | HFC | Coil is leveled, formed, welded 4 convex-concave rolls | 1960 | No | 2,500 psi Max | 2 to 12-3/4 | 0.154 to 0.375 | A,B | 45 ft Max | Ultrasonic inspection |
| Stelpipe Ltd., Welland, Ontario, Canada | Page-Hersey Facility | HFC Thermatool | Coils are leveled and edge trimmed. preforming is done in 4 convex-concave rolls followed by a series of 3 vertical rolls and 3 fin-pass rolls. Welding, flash-trimming, seam-annealing, sizing, rotary cut-off | Replaced original Page-Hersey mill built in 1950 | Yes (Hydraulic for 8-inch and up) | | 2⅜ to 16<br>4½ to 8 STD<br>8 to 16 XH DRL & SRL | 0.083 to 0.500 | A through X65 | 21 to 42 ft | 100% ultrasonic inspection of seam after annealing and again after hydrostatic test. All skelp from Stelco. |
| | ERW/SRM 150,000 tons/yr | HFI 600 kw | 5-inch mother shell made from coils. Coil is leveled, edge-conditioned, straight-edge formed, welded, flash-trimmed, cooled, sized, straightened, cut to 200 ft. lengths. 5-inch tube is induction heated, stretch reduced, cut-off, cooled, straightened | Began production in 1993, replaced CW mill | No | API Minimum | 0.840 to 4½+H5<br>½ to 4 | 0.0109 to 0.337 | B | 21 to 42 ft | 100% seam inspection by ultrasonics after welding. 100% full-body eddy-current inspection after hydrostatic test |
| Stupp Corporation, Baton Rouge, LA | Thomas Rd Facility 400,000 tons/yr | HFC 600 kw thermatool 4 Cheston seam annealers | Coil is leveled, edges rotary trimmed, preform rolls and cage-forming to U-shape, 3-stand fin pass, 5 seam-guide rolls, welding, flash-trimming, air/water cooling, sizing and straightening, rotary cut-off | Began production in 1970 replacing a nearby mill which was then shut down | No | 100% SMYS for 20 sec. 2 testers 3000 and 5000 psi | 8⅝ to 24 | 0.188 to 0.500 | B through X70 | up to 105 ft | 100% electromagnetic seam inspection after sizing and straightening. 100% ultrasonic seam inspection after hydrostatic test |
| Tex-Tube Company, Houston, TX | 1 Mill | HFC(ERW HF-with contacts)-- Thermatool-400 kw | Slit coil is leveled, formed through cage forming, welded, flash removed, seam-annealed, air cooled, water cooled, sized, cut-off, beveled, ID flashed removed via high pressure water flush, hydrotested | Mill installed in 1956 to replace previous mill installed 1951. Mill rebuilt in 1979 to produce 8-5/8"OD pipe | No | 100% SMYS for 5 sec. for stock, 10 sec by customer request | 3½ to 8⅝ | 0.109 to 0.322 | B through X60 | Up to 60 ft | GMI and ultrasonic weld line inspection prior to cutting to length. Full body GMI after hydrotest. |
| | 2 Mill | HFC Thermatool-300 kw | Slit coil is leveled, formed via conventional forming, welded, flash trimmed, seam-annealed, air cooled, water cooled, sized, cut-off, beveled, ID trim is flashed out with high pressure water, hydrotested | Mill installed in 1960 | | 100% SMYS for 5 sec., 10 sec. by customer request | 2⅜ to 4½ | 0.109 to 0.250 | B through X60 | up to 45 ft. | GMI and weld line ultrasonic prior to cutting to length |
| Tubacero SA, Monterrey, Mexico | Mill No. 2 | | | Began operating in 1943 | | | 18 through 24 | .0156 to 0.500 | | 51 ft | |
| | Mill No. 3 | | | Began operating in 1943 | | | 6⅝ through 16 | 0.156 to 0.500 | | 44 ft | |
| Tuberia Laguna SA de CV, Gomez, Palacio, Mexico | | HFI | | | No | | 6⅝ through 24 | 0.250 to 0.375 | up to 60X | | |
| U.S. Steel | Lorain Facility, Lorain, OH | HFC 280 kw 450 kHz | Coils are leveled, edges-trimmed, skived, formed by 4 breakdown/cage rolls, 3 fin-pass stands, welded, flash-trimmed seam-normalized (1700F), air/water cooled sized, straightened, cut off | Began operations Jan 1967 | | | 4½ to 6⅝ | 0.156 to 0.280 | A through X70 | 45 ft max. | 100% seam inspection by ultrasonics after welding. 100% sam inspection by electromatnetics after hydrostatic test |
| | McKeesport, PA (See Camp Hill Corp.) | | | | | | | | | | |
| Villacero Tuberia Nacional SA de CV Nuevo Leon, Mexico | | HFI | Slit coil is leveled, edge trimmed formed by breakdown rolls, welded, cut to length, and full-body normalized or stretch reduced | | | | 0.405 to 6.625 | 0.068 to 0.258 | A,B | | |
| Welland Pipe, Ltd., Ontario, Canada | See Stelpipe Ltd. | | | | | | | | | | |

(a) HFC refers to high-frequency welder with sliding contacts
(b) HFI refers to high-frequency welding by induction (no contact)
(c) SRM refers to stretch-reduction mill

Table C-7  Past Manufacturers of Submerged-Arc Welded Pipe

| Manufacturer | Mill/ Capacity | Key Manufacturing Steps | Key Dates | Cold Expanded Yes/No | Hydrotest Info | Diameter Range, inch | Wall Thickness Range, inch | Grades | Lengths | Other Information |
|---|---|---|---|---|---|---|---|---|---|---|
| Ameron (American Pipe and Construction Co.) | Portland, OR | | 1965 to 1971 | No | | 24 to 44 | | | | Special applications(a) |
| Armco Steel Corp | Houston, TX | Cans formed by U-ing and O-ing. Welds started at each end and over lapped at mid-length | About 1967 to about 1980. Acquired from A.O. Smith Corp which had converted the facility from flash-welding. | Yes mechanical | API Minimum | 24 to 36 | 0.250 to 0.750 | B through X70 | 40 ft. | Succeeded A.O. Smith's flash weld process at their Houston facility |
| Bethlehem Steel Corp., Steelton, PA | | 20-ft plates roll-formed and submerged-arc seam welded. Double jointed to obtain 40-foot joints | Began operating in 1955 replaced in 1960 by current UOE mill | Yes hydraulic | API Minimum | | | | 40 ft after double-jointing | Carbon arc welding of low pressure pipe, 1932. Sub-arc welding began in 1940 |
| A.M. Byers Company | | Large OD (greater than 14-inch OD) made by cold-forming and fusion welding. Material was wrought iron | | No | | | | | 16 to 22 ft | Also used hammer weld. Special applications(a) |
| Chemtron Corp (Formerly tube Turns, Inc.) | Louisville, KY | Normalized and stress relieved | 1971 to 1983 | No | | 20 to 70 | 0.375 to 3.500 | up to X60 | | Special applications(a) |
| Claymont Steel Corp | Claymont, DE | Not Available | About 1951 through 1970 | Unknown | Unknown | 22 to 36 | Unknown | Unknown | Unknown | Under various names Claymont, Wickwire-Spencer, 1954-1960, a division of Colorado fuel & Iron Corp. Acquired by Phoenix Steel Corp between 1960 and 1964 |
| Gulf & Western Mfg Co (Formerly Taylor Forge, Inc.) | Chicago, IL | Normalized and stress relieved | 1940 to 1978 | No | 16 to 60 | 0.375 to 1.500 | up to X60 | | | Special applications(a) |
| Kaiser Steel Corporation | Napa, CA 540,000 tons/yr | Edges of plate planed, beveled, crimped. Cans formed by U-ing and O-ing. Inside weld first, then outside weld. Hydraulically expanded, hydrotested | | Yes, hydraulic | API Minimum | 18 to 42 | | B through X70 | 40 ft | |
| Kane Industries, Inc (Formerly Kane Boiler Works, Inc.) | Galveston, TX | Stress relieved | 1971 to 1987 | No | | 20 to 96 | 0.250 to 1.500 | up to X65 | | Special applications(a) |
| Republic Steel Corporation | Gasden, AL | Plate ends sheared, edges planed and shot-blasted. Edges crimped in a series of rolls. Cans formed by pyramid rolls. Cans tack-welded end-to-end by short tack welds on each side of opening. Welding cage forces edges together for OD welding. 200F preheat by gas burners. Water-cooled copper back-up bar. After OD welding tack welds between ends of cans were cut, tabs were welded on for ID welding. ID welding followed by inspection, mechanically expansion, hydrotesting, beveling, inspection | Began in 1951, discontinued in the 1970's | Yes, mechanical | API Minimum 2,500 psi Maximum | 20 to 30 | 0.252 to 0.500 | A through X70 | 27 to 3.5 ft | Open hearth steel made and plates rolled on-site |
| Smith Industries Inc | Houston, TX | | 1973 to 1982 | No | | up to 40 | up to 0.750 | up to X60 | | Special applications(a) |
| Teledyne Pipe | Galveston, TX | | 1975 to 1985 | No | | 20 to 84 | 0.375 to 3.000 | | | Special applications(a) |
| Texas Pipe Bending Co | Houston, TX | | 1980 to 1984 | No | | 14 to 64 | up to 2.750 | | | Special applications(a) |
| U.S. Steel Corporation and Subsidiaries | National Tube Company, Christ Park, PA Works | Plates planed, formed into cans by pyramid rolls, welded with single pass submerged arc process from outside onto back-up shoe. Approximately 500 miles made. | Began operating in 1930, ceased operating in 1932 | No | Unknown | 14 to 30 | Unknown | Unknown | 30 ft. | |
| Consolidated Western Steel Corp Originally Western Pipe | | Plate, obtained from Columbia Steel (Geneva, UT) or Kaiser Steel (Fontana, CA) were squared, planed, and beveled 5% so inner edges of can would touch. Outer corners were chamfered to guide flexible weld head. Edges were crimped | Began operating in 1946. Closed down or moved all facilities to Provo, UT and Orange, TX by 1956 | Yes, hydraulic | API Minimum | 24 to 36 | Unknown | Unknown | 28 to 31 ft | Western Pipe and Steel Company had an API certificate as early as 1940. By the time, U.S.S bought Consolidated Western (1948) the following plants were operating |

Table C-7. Past Manufacturers of Submerged-Arc Welded Pipe

| Manufacturer | Mill/ Capacity | Key Manufacturing Steps | Key Dates | Cold Expanded Yes/No | Hydrotest Info | Diameter Range, inch | Wall Thickness Range, inch | Grades | Lengths | Other Information |
|---|---|---|---|---|---|---|---|---|---|---|
| | and Steel Co. Purchased by U.S.S. in 1948 | by rolls and cans formed in pyramid rolls OD weld was made first as can edges were held by cage rolls Each OD weld had to be hand welded at the end by hand-held SAW devices because cage rolls could not hold the ends accurately These end welds were called "squirt" welds Tabs were welded on each end to start the ID weld and allow it to run off Welds were visually inspected The ends were mechanically belled to accommodate the expander plugs The pipes were then hydraulically expanded, and beveled | | | | | | | | Maywood (Los Angeles) -- 540,000 tons/yr. First mill to make 30-inch OD pipe (1946) Vernon (Los Angeles) -- 67,000 tons/yr South Francisco --246,000 tons/yr. Berkeley -- 30,500 tons/yr Fresno -- 4,500 tons/yr Phoenix (Phoenix, AZ) -- 2,500 tons/yr Orange (Orange, TX) -- 300,000 tons/yr Only Maywood, Orange, and Berkeley remained open by 1954 By 1956 all operations had been moved to Provo, UT and Orange, TX. In 1955, the Orange Plant became the American Bride Div of U.S.S. and the Provo Plant became the Consolidated Western Div of U.S.S |
| National Tube Company, National Works, Mckeesport, PA | | Plates obtained from Homestead and later Edgar Thompson Works Edges planed, beveled, and crimped Cans formed by U-ing and O-ing OD weld made first as can was held by cage rolls Ends squirt welded by hand Tabs welded on for ID weld start and run-off ID-welded, inspected, hydraulically expanded, hydrotested, and beveled | Began operating March, 1950 Operation ceased in ---- | Yes, hydraulic until ----, then mechanical | API Minimum | Initially 20 to 36 later 24 to 36 | Initially 0 250 to 0 500 later to 0 656 | Initially to X52 later to X65 | 40 ft | By mid 1960s seams were inspected by fluoroscopy and ends were film x-rayed |
| American Bride Div., Orange, TX | | Plates formed into cans by U-ing and O-ing OD weld made 1st squirt welds | Began as Consolidated Western facility, 1951, became Am Bridge in 1955 ceased operation by 1978 | Yes | API Minimum | Initially to 36 later 24 to 42 | 0 250 to 0 750 | Initially to X52 later to X65 | 40 ft | |
| Consolidated Western Div., Provo, UT | | Plates obtained from Columbia-Geneva Steel next door were planed and formed into cans by U-ing and O-ing without initial edge crimping ID weld made first while can held stationary in hydraulic clamps OD weld made last, no squirt welds needed Hydraulically expanded, hydrotested, beveled and inspected | Began operation in 1956, ceased operating in ---- | Yes, hydraulic | API Minimum | 20 to 40 | 0 250 to 0 625 | Initially to X52, later to X65 | 40 ft | |
| Texas Works, Baytown, TX 500,000 tons/yr | | Plates edge milled, beveled, crimped Cans formed by U-ing and O-ing Cans tack-welded and tabs are attached for run-on, run-off ID welded (3-wire), OD welded (3-wire), inspected, mechanically expanded, hydrostatically tested, end-beveled, inspected | Began operating in 1977 Facility sold to SAW Pipes USA, Inc in 1993 | Yes, mechanical | API Minimum | 24 to 48 | 0 250 to 1 000 | A through X70 | 40 ft | Electric furnace, continues cast steel with desulfurization available Seams inspected by fluoroscopy, film X-rays at ends and ultrasonics |

(a) These facilities probably made short runs of pipe for special purposes such as compressor stations and for fittings and attachments

## Table C-8. Current Manufacturers of Submerged-Arc Welded Pipe

| Manufacturer | Mill/ Capacity | Key Manufacturing Steps | Key Dates | Cold Expanded Yes/No | Hydrotest Info | Diameter Range, inch | Wall Thickness Range, inch | Grades | Lengths | Non Destructive Inspection and Quality Factors |
|---|---|---|---|---|---|---|---|---|---|---|
| Berg Steel Pipe Corporation, Panama City, Florida | 120,000 tons/yr non-UOE | Plates edge planed, cans formed using pyramid rolls, can edges crimped, hydraulically closed and continuously tacked by a one-wire mig weld from the outside. Inside weld made next with tack weld as backup by 4-wire SAW process Tabs used for run-on and run-off. Final outside weld made with 5-wire SAW process, consumes all tack weld, burns into inside weld. Sizing, end facing, hydrotesting, final inspection and stenciling. | Began operating in 1980 | No | API Minimum | 24 to 64 | 0.250 to 1.5 | B through X70 | 40 ft, double-jointing to 80 ft | Ends of welds fluoroscoped after welding, seam ultrasonically inspected after hydrotesting, ends film X-rayed prior to final visual inspection, visual inspections performed after welding are as final step. 12 worldwide plate suppliers |
| Bethlehem Steel Corp., Steelton, PA | 300,000 tons/yr UOE | Plates side and end planed, edges crimped, U-ing, O-ing to form cans, inside weld made first, pipes butted end-to-end permits continuous welding. Outside welding follows also continuous from one pipe to the next. Inspection after welding, hydraulic expansion, hydrotesting, weld inspection, end facing, final inspection | Began operating in 1960 | Yes hydraulic | API Minimum | 20 to 42 | 0.281 to 0.875 | B through X70 | 40 ft, double-jointing to 80 ft | Plate from Burns Harbor or Sparrows Point facilities. Desulfurization, controlled-rolling. Key in top of O-ing press assures true alignment of can edges, cans gradually closed and held by cage rolls as ID weld is deposited |
| Napa Pipe Corporation | UOE | Plates from Portland or Fontana facilities are sheared, beveled, planed, and crimped. Cans formed by U-ing and O-ing. Cans washed and tacked. Lead-in and run-off tabs attached ID weld made by boom guided by bevel. OD welded, fluoroscoped. hydraulically expanded, hydrotested, end-faced, inspected | Began operating in 1987 (formerly a Kaiser facility) | Yes, hydraulic | API Minimum | 16 to 42 | 0.281 to 1.000 | B through X70 | 40 ft, double-jointing to 80 ft | Complete fluoroscopy of seam after welding. Full body ultrasonically inspected after hydrotesting and end-facing. Final visual inspection plus film radiography of ends of seams |
| Productora Mexicana de Tuberia S.A. de C.V., Michoacan, Mexico | 300,000 metric tons/yr UOE | Plate edges trimmed, crimped. Cans formed by U-ing and O-ing, cans cleaned, MIG-tack welded from OD side, fitted with run-on, run-off tabs ID welded, OD welded, tabs removed, inspected, cleaned, mechanically expanded, cleaned again, hydrotested, inspected, end-faced, inspected | Began operating in 1986 by Mexican Gov't Privatized in 1992 | Yes, mechanical | API Minimum | 16 to 48 | 0.250 to 1.125 | Primarily X60 through X80 | 40 ft | Plates ultrasonically inspected. Film X-ray of end of welds and ultrasonic inspection of seam after welding and again after hydrotesting. Japanese technology. Currently owned 51% by IMEXA, 49% by TUBACERO. |
| SAW Pipes USA, Inc., Baytown, TX | 500,000 tons/yr UOE | Cans edged milled, beveled and crimped. Cans formed by U-ing and O-ing. MIG tack welded and fitted with run-on and run-off tabs ID welded (3-wire), OD welded (3-wire) (tack-weld consumed) fluoroscoped, mechanically expanded, hydrotested, end-faced, inspected | Acquired from U S Steel in 1993 | Yes, mechanical | API Minimum 3000 psi maximum | 24 to 48 | 0.250 to 1.000 | Primarily X60 through X80 | 40 ft, double-jointed to 80 ft | Fluoroscopy of entire seam after welding and again after hydrotest Entire seam ultrasonically inspected after hydrotesting weld ends film X-rayed. |
| Stelpipe, Welland, Ontario and Camrose, Alberta, Canada | UOE | Plates shot blasted, edge trimmed, crimped Cans formed by U-ing and O-ing. Can edges tack welded manually. Ends milled. Run-on and run off tabs attached to ends ID welded by travelling boom (stationary pipe), OD welded by stationary welded (pipe moving) Pipes tipped to remove foreign matter, hydraulically expanded, hydrotested, fluoroscoped, end-faced | | Yes, hydraulic | API Minimum | 20 to 36 at Welland 20 to 42 at Camrose | 0.252 to 0.625 | A through X70 | 40 ft | Fluoroscopy to entire seam after hydrotesting |
| Talleres Acero Rey, S A de C V Monterrey, Mexico | 45,000 t/yr, UOE bending process, straight seam | Trimming and crumping of plate edges. U-ing, O-ing, tack welding, inside and outside welding, U T and X-raying testing of weld, mechanical expanding. Hydrostatic testing, chamfering of pipe ends, final inspection | Began producing in 1985 API certified in 1991 | Yes, mechanical and hydraulic for 18 in and up | API minimum 6000 psi max | 18 to 120 | 0.250 to 3.375 | API 5L Gr B, X42 to X80, ASME, ASTM, AWWA | up to 50 ft | 100% ultrasonic inspection of seam after hydrostatic test. 100% X-ray test of beams before final inspection |

Table C-8. Current Manufacturers of Submerged-Arc Welded Pipe

| Manufacturer | Mill/ Capacity | Key Manufacturing Steps | Key Dates | Cold Expanded Yes/No | Hydrotest Info | Diameter Range, inch | Wall Thickness Range, inch | Grades | Lengths | Non Destructive Inspection and Quality Factors |
|---|---|---|---|---|---|---|---|---|---|---|
| Tubacero, S.A de C.V. Monterrey, Nuevo Leon, Mexico | Mill No. 1 | Continuously forms steel plate that is tack-welded by the ERW process | | | | 20 to 48 | 0.219 to 1.25 | up to X80 | 51 ft | |
| | Mill No 2 | Continuously forms steel plate that is tack-welded by the ERW process | | | | 18 to 36 | 0.219 to 1.125 | up to X80 | 51 ft | |
| | Mill No 5 | Pyramid forming process (non API Grades) | | | | 20 to 150 | 2.5 | | 10 to 28 ft | |
| | 350,000 metric tons | PMT, and longitudinally welded formed by the U-ing and O-ing process | (See Productora Mexicana de Tuberia for more details) | | | 16 to 48 | 1.125 | | | |

## Table C-9. Current and Past Manufacturers of Spiral-Welded Pipe.

| Manufacturer | Mill/ Capacity | Key Manufacturing Steps | Key Dates | Cold Expanded Yes/No | Hydrotest Info | Diameter Range, inch | Wall Thickness Range, inch | Grades | Lengths | Non Destructive Inspection and Quality Factors |
|---|---|---|---|---|---|---|---|---|---|---|
| Armco Steel Corp., Middletown, OH | | Submerged arc welded | Began in 1961. No longer producing line pipe. | | | up to 36 | | | | |
| Ipsco Inc., Regina, Sask. and Edmonton, Alberta | 200,000 t/yr | Made from coil, double submerged-arc-welded | First mill in 1967, 2nd mill in 1969, 3rd mill in 1972. CURRENT | Unknown | API Minimum | 16 to 80 | 0.156 to 0.600 | up to X80 | | Ultrasonic inspection |
| Lone Star Steel, Dallas, TX | | Double-submerged-arc welded | 1961 to early 80's. No longer producing line pipe. | Unknown | API Minimum | 24 to 84 | 0.250 to 0.500 | up to X46 | | |
| Stelpipe, Welland, Ontario, Canada | | Plates cleaned and shot blasted. Plate ends and edges squared and beveled. Joined end-to-end up to 160 ft by stationary cross-seam welder, then joined to continuous strip by traveling cross seam welder. Spiral forming mill preforms edges and spirals he strip, continuously GMA tack-welding the seam. 80-ft lengths are then cut. Cross-seam welds are completed. Then ID spiral 2-wire weld is made followed by OD spiral 2-wire weld spiral and cross-seam welds fluoroscoped. Pipes are mechanically expanded, hydrotested, inspected, and beveled. | CURRENT | Yes, mechanical with spiral relief groove | API Minimum | 36 to 60 | 0.252 to 1.125 | | up to 80 feet | Fluoroscopy of spiral and cross-seams after welding and again after hydrotesting. Film X-rays of T-welds. Ultrasonic spiral and cross-seam welding after hydrotesting. |
| Taylor Forge, Inc., Chicago, Il | | ERW | In operation in the 1950's. No longer producing line pipe. | | | | | | | |
| Tubesa, S.A. de C.V., San Louis, Potasi, Mexico | 80,000 t/yr | Double-submerged arc-welded | CURRENT | Unknown | API Minimum | 20 to 120 | 0.312 to 1.000 | | | Ultrasonic inspection |

# 8.0 IDENTIFYING UNKNOWN SAMPLES OF LINE PIPE

Occasionally, a pipeline operator encounters a pipeline or a segment of a pipeline for which the history of manufacturing may be unknown. Aside from the diameter which can be easily measured and the wall thickness which can be determined ultrasonically, how does one go about identifying the grade, the material (iron or steel), the seam type (if any), the manufacturer, and whether or not the material is API line pipe? Some helpful techniques are discussed below. First, it is wise to thoroughly check one's files. Sometimes the mill certificates for old pipe orders still exist. Alternatively, if a third-party inspection company was used, it may be worthwhile contacting that company to see if they still have the mill certificates.

## Die Stamped or Stenciled Information

Until 1954 metal die stamping was required to be placed on each piece of API line pipe unless otherwise agreed between the manufacturer and purchaser. Within one foot of either end of the pipe these marks would be stamped on the OD surface in ¼-inch high letters. Until 1942, the die-stamped information included the manufacturer's mark, the API Monogram and the grade of pipe (A, B, or C) if seamless or electric welded. After 1942, die-stamped marks were to include the class of pipe, F for butt-welded, L for lap-welded, E for electric-welded or S for seamless and the type of material, I for open hearth iron, WI for wrought iron, and R for rephosphorized Class II steel.

Starting in 1954, the die stamping was not required for pipe of 14-inch OD and larger and by 1969, its use was so restricted that it is doubtful that much die-stamping was done thereafter.

Paint stenciling of some information was done in addition to die stamping and all information was paint-stencilled once die stamping was eliminated. In larger pipe sizes, the paint stenciling is or was often done on the ID surface. Early paint stenciling (1928-1942) included the length of the pipe, the weight of the pipe, and the test pressure. After 1942, it included the size, weight, grade, class, material, and length. Test pressure was included only if a higher-than-normal value was applied.

Presumably, one has a chance of finding the outside surface die-stamp or paint stencil marks if the coating is removed carefully. If the marks were put on the ID surface, they would be useless for identifying the pipe unless it was taken out of service. In that case, other, more positive steps could be taken to identify the pipe.

### Seam Welds and Other OD Physical Features

External physical characteristics can be somewhat helpful. Submerged-arc seam welds and flash-welds are easy to spot on an in-service pipeline. The rounded crown of an SAW seam at least narrows the range of manufacturers. The large square flash of an A.O. Smith flash weld is unique. While some ERW welds exhibit an OD flash, none is as wide or as prominent as the A.O. Smith weld. Very old A.O. Smith pipe (before 1930) had a very wide shielded metal-arc welded crown. It is a very high and wide crown that could not easily be mistaken for an SAW seam.

Early National Tube (McKeesport) DSAW pipe and early consolidated Western DSAW pipe is recognized by the "squirt" welds at ends of the pipe. At these locations, the OD bead is abruptly wider for about 3 to 8 inches from each end.

At times in the past, most manufacturers embossed their brand on the pipe, especially on lap welded pipe during hot forming. This was typically done on lap-welded pipe by means of an engraved roll in the pullout stand after the welding furnace.

"In addition, when most CW pipe skelp was rolled on skelp mills it was common to use an engraved roll in the last stand of the mill, resulting in a brand located 180 degrees from the weld. In both the lap weld and CW cases, the name or brand recurred at an interval reflecting the diameter of the engraved roll. In addition, it was common to include additional information in the rolled-in brand including steel type (open hearth or Bessemer) and, in some cases, year and even quarter of the pipe manufacture."[8-1] Lap-welded pipe was often "knurled" or "spellerized". Spellerizing was a waffle-like pattern embossed on the surface. Occasionally lap-welding was so poorly done that the outer edge of the bondline is visible.

Some older ERW materials were flash-trimmed at the OD surface in a manner which makes the weld visible. One such was the Youngstown Sheet and Tube Company process between 1940 and 1950. During at least part of that period the OD flash was trimmed by a tool so wide that it created a flat spot ½- to ¾-inch wide along the seam. Pipes trimmed in this manner are probably identifiable as Youngstown pipe.

## Material Sampling

If it is possible to destructively examine the material by cutting a sample, much can be learned about the material. With a very small sample such as a coupon from a 2-inch hot tap one can determine the chemical content and whether or not the material is iron or steel. With a ring of the material one can determine the type of seam weld, if any, and the tensile properties and fracture toughness properties. It is possible to distinguish high-frequency ERW seams from low-frequency ERW seams by the microstructure. By testing a significant number of randomly selected samples, one can establish the minimum yield strength for a pipeline of unknown material. With this information and knowing the type of seam, the pipeline operator should be able to establish a maximum allowable operating pressure for a pipeline comprised of an unknown line pipe material. Of course, it may be necessary to hydrostatically test the pipeline to 1.25 times that pressure to be able to utilize it.

## Determining Missing Information
## When Some Information is Known

Often the pipeline operator knows some information such as the age of the pipe or the manufacture or both, but does not know the yield strength or the seam type. It is often possible to determine missing information when some information is known. One item that helps is the summary of API 5L pipe materials (not the X grades) shown in Table 8-1. This table permits one to use some information about API line pipe. It is emphasized that we are talking about API line pipe. If it is suspected that the material is not API line pipe, then this table would be useless.

Examples of the use of Table 8-1 are as follows. Suppose one has a pipe material with an obvious flash-weld or SAW seam. The minimum yield strength of the material can be no lower than 30,000 psi because these materials were not manufactured from any of the iron or Bessemer steel groups. Suppose one has an ERW material known to have been manufactured before 1969. This material is probably of Grade A or higher (30,000 psi minimum yield). The only doubt arises because of the short time window 1942-1945 when open hearth iron was permitted for the manufacture of ERW pipe. If that possibility cannot be ruled out by other means, a hot tap coupon could be removed to show whether the material is iron or steel.

Another method of determining information involves reviewing what the various manufacturers made. This can be done

on the basis of Table M-1 through M-9.  A list of examples
follows.

*American*.  American Steel Pipe began making line pipe
in 1963.  They have made only ERW pipe of Grade B or higher
strength materials.  All of their ERW pipe has been made with
high-frequency welders.  Their Mill No. 1 was started in 1963 and
turns out 10.75- to 20-inch OD pipe.  Their Mill No. 2 was
started in 1989, and it produces pipes ranging from 16- to
24-inch OD.  So American pipe is at least Grade B material and it
is high-frequency welded ERW.

*A.O. Smith*.  A.O. Smith Corporation made only flash-
welded steel pipe in the period between 1930 and 1969.  All of it
would have been at least Grade A material.  Between 1969 and 1973
they produced DSAW pipe in diameters up to and including 36-inch.
Their DSAW welds were unique in that they were started at each
end and met in the middle of the pipe length.  They were no
longer certified to make API line pipe after 1973.

*Bethlehem*.  Bethlehem Steel Company began making ERW
pipe in 1957 with a low-frequency mill in sizes ranging from
5.5625- to 16-inch OD with a minimum strength of Grade A.
Bethlehem's high-frequency mills were as follows:

1963-1982       Grades A through X52     2,375- to 6.625-inch OD
1970-1982       Grades A through X60     5.562- to 16-inch OD.

*Geneva*.  Geneva Steel in 1991 made only high-frequency
ERW of at least Grade B material.

*Interlake*.  Interlake Steel made low-frequency welded
ERW in sizes 4.5- through 8.625-inch OD in materials of Grades A
through X52 prior to 1980.

*J&L*.  Jones & Laughlin Steel Corporation made both
seamless and ERW line pipe at their Aliquippa works.  The grades
available ranged from Grade A through Grade X60.  J&L's low-

frequency ERW mill produced pipe between 1957 and 1964 in the
diameter range 6.625- to 12.75-inch OD. In 1965 the low-
frequency mill was replaced by a high-frequency mill making pipe
in sizes ranging from 4.5- through 12.75-inch OD. In 1985 after
J&L had been absorbed by LTV Corporation, the Aliquippa pipe
mills were closed.

*Kaiser.* Kaiser made low-frequency ERW pipe at their
Fontana, California mill from 1950 through 1964 in sizes ranging
from 4.5- through 16-inch OD, Grades A through X52. This mill
was converted to a high-frequency mill in 1965 making the same
sizes and grades. Kaiser operated a second ERW mill at Napa,
California between 1950 and 1964. The latter was a low-frequency
mill and made pipe in sizes ranging from 6.625- through 20-inch
OD in Grades A through X52.

*Lone Star.* Lone Star Steel began producing ERW line
pipe in 1953. Their Mill No. 1 produced low-frequency ERW pipe
in sizes ranging from 6.625- through 16-inch OD, Grades A through
X52. The welder in this mill was replaced in 1969 by a high-
frequency welder. The same materials were made although
currently the minimum size is 8.625-inch OD and the grades
include Grade B through X65. Lone Star's Mill No. 2 was started
in 1953. It produced low-frequency ERW in sizes ranging from 1.9-
to 6.625-inch OD, Grades A through X52. In 1963 the welder in
this mill was changed to a high-frequency-induction welder which
was altered in the mid 70s to a sliding contact welder. In 1980
the welder was replaced again with an entirely new high-frequency
welder. Currently the minimize size made in Mill No. 2 is
2.375-inch OD and the minimum grade is Grade B.

*LTV.* LTV Steel Tubular Products took over the
operating ERW mills of Republic Steel Corp. in 1986. These mills
make high-frequency ERW pipe materials ranging in sizes from
2.375- to 16-inch OD in Grades B through X65.

*National Tube*. National Tube Company produced both lap-welded and seamless pipe materials prior to World War II. By 1943, however, they had ceased making lap-welded pipe. All National Tube seamless pipe made prior to 1943 has a minimum yield strength of 30,000 psi (Grade A). Between 1946 and 1964, National Tube products could only have been seamless pipe and the minimum grade would have been Grade A.

*Newport*. In 1950 Newport made low-frequency ERW pipe in sizes ranging from 4.5- to 8.625-inch OD, Grades A through X52 (Newport became Interlake). By 1983 when Newport was revived, they made high-frequency ERW pipe on one of two mills which are currently in operation.

| 8-inch Mill | Grades B through X52 | 4.5 to 8.625-inch OD |
| 16-inch Mill | Grades B through X52 | 8.625 to 16-inch OD. |

*Republic*. From 1929 until 1961, Republic made low-frequency ERW pipe in one of three mills in sizes ranging from 2.375 through 16-inch OD. The materials ranged from Grade A through X52. In 1961, they converted these mills to high-frequency welders making the same materials. In 1963, they installed their Mill No. 5 which still exists today as an LTV mill. The minimum grade was Grade A in 1963 but currently is Grade B.

*Stupp*. Stupp made only high-frequency ERW pipe. Their size range was and is 8.625- through 24-inch OD. They claim to make all API 5L Grades through X80. The minimum grade would have been and may still be Grade A although it would be hard to find Grade A skelp today.

*Tex Tube*. Tex Tube has produced ERW pipe since 1951 and API line pipe since 1954. Their early mill must have been low-frequency. Their current mills (2) are high-frequency. The maximum size they produce is 8.625-inch.  The minimum grade of line pipe that they could have produced would have been Grade A.

*U.S. Steel*.  In the sizes below 24-inch, all U.S.Steel pipe would be either seamless pipe or high-frequency ERW pipe. The minimum grade would be Grade A.

*Youngstown*.  Youngstown Sheet and Tube Company produced only seamless and d.c. ERW pipe.  The minimum grade would have been Grade A.  Sizes of ERW pipe could have ranged from 6⅝ to 26-inch although 22-inch was the largest size made after 1945. The seamless pipe ranged from 2⅜ to 14-inch.  The d.c. ERW pipe was made at Brier Hill works in Youngstown Ohio.  The Brier Hill works was closed about 1978.

## Reference

8-1. E. A. Jonas', private communication, 1996.

Table 8-1. API 5L Line Pipe

| | Butt-Welded Pipe | Lap-Welded Pipe | Seamless Pipe | Flash-Welded Pipe | ERW Pipe | DSAW Pipe |
|---|---|---|---|---|---|---|
| | <6⅝ OD | | | | | ≥18" OD |
| Wrought Iron 24,000 psi | Discontinued by Mar. 1962 | Discontinued by Mar. 1963 | — | — | — | — |
| Open-Hearth Iron 24,000 psi | Discontinued by Jan. 1961 | Discontinued by Jan. 1961 | Started June, 1932 Discontinued by Jan. 1961 | — | Started Nov. 1942. Discontinued by Nov. 1945 | — |
| Pre A25 Steel Class I 25,000 psi | Discontinued by Apr. 1969 | Discontinued by Mar. 1962 | — | — | — | — |
| Pre A25 Steel Class II 28,000 psi | Discontinued by Apr. 1969 | Discontinued by Mar. 1962 | — | — | — | — |
| Bessemer Steel 30,000 psi | Discontined by Apr. 1969 | Discontinued by Mar. 1962 | — | — | — | — |
| Grade A25 Steel 25,000 psi | Started Apr. 1969 | — | Started Apr. 1969 | — | Started Apr. 1969 | — |
| Grade A Steel 30,000 psi | — | — | Started Jan. 1928 | Started Mar. 1955* | Started May 1942 | Started May 1962 |
| Grade B Steel 35,000 psi | — | — | Started Jan. 1934 | Started Mar. 1955* | Started Jan. 1934 | Started May 1962 |
| Grade B Steel 38,000 psi | — | — | Started Jan. 1930. Discontinued by Jan. 1934 | — | Started July 1931 Discontinued by Jan. 1934 | — |
| Grade B Steel 40,000 psi | — | — | Started Jan. 1928. Discontinued by Jan. 1930 | — | — | — |
| Grade C Steel 45,000 psi | — | — | Started Jan. 1928. Discontinued by Mar. 1955 | — | Started July 1931 Discontinued by Mar. 1955 | — |

* Although A. O. Smith made flash welded pipe as early as 1930, they did not have an API license until the 14th Edition of API 5L (March, 1955).

# SECTION D

# API LINE PIPE SPECIFICATIONS

# 9.0 INTRODUCTION

The American Petroleum Institute has published and updated specifications for the manufacturing of line pipe since 1928. The first such specification, API Standard No. 5-L, First Edition, was issued in January, 1928. The First Edition was a "pocket-sized" (4 x 7 inches) 38 page document. The most recent edition: API Specification 5L, Forty-First Edition, is dated April 1, 1995. It is an 8½- by 11-inch book of 119 pages. Along the way the document has undergone many changes but the essential purpose has remained the same, namely, "to provide standards for pipe suitable for use in conveying, gas, water, and oil in both the oil and natural gas industries". The elements of pipe manufacturing addressed by the standard have remained essentially the same, namely, processes of manufacturing, material properties, dimensional requirements, workmanship, inspection, testing, threading and couplings, and marking and documentation. Nevertheless, if one were to compare the First Edition to the Forty-First Edition, one would notice significant modifications mostly in response to both increasing emphasis on product quality and improvements in manufacturing technology. Summaries of the requirements of each API 5L edition are listed in Table 9-1, and summaries for API 5LX editions are listed in Table 9-2. Listings of which manufacturers held 5L and 5LX licenses and when they held them are presented in Tables 9-3, 9-4, and 9-5. The contents of the various editions are described below.

## Early API 5L Specifications (1928-1948)

### The First and Second Editions

When the first API specification was issued in January, 1928, it was titled A.P.I. Standard No. 5-L, First Edition, A.P.I. Line Pipe Specifications. It consisted of 38 pages in a pocket-sized 4- by 7-inch format. It listed five materials, Bessemer steel, open hearth steel, electric steel, wrought iron and open hearth iron. Three processes for manufacturing line pipe were considered: butt-welded pipe (nominal size ⅛-inch through 3-inches), lap-welded pipe (nominal size 1¼ inches through 20-inch outside diameter) and seamless (nominal size 1¼ inches through 20-inch outside diameter). It was implied that larger-diameter pipe could be made as API line pipe by agreement between the purchaser and the manufacturer. Butt-welded and lap-welded pipe could be made from iron or steel but seamless pipe was to be made only from steel. No mention of electric welded seam processes was made in the First Edition although flash-welded pipe was available at the time. Neither is hammer-welded pipe mentioned; in fact hammer-welded pipe is never mentioned in any API line pipe specification.

The chemical compositions for steel pipe listed in the First Edition are shown in Table 9-1. Carbon is not mentioned so there was no limit on carbon at that time. Manganese was limited by both a maximum and a minimum value. Phosphorus was limited in most cases by a maximum value, but a minimum value existed as well for rephosphorized steel. Sulfur was limited by a maximum value. The values were given for a ladle analysis, and no variations for check analyses were listed. The chemistry for open-hearth iron was limited to 0.16 percent by weight for the sum of the carbon, sulfur, phosphorus, silicon, copper and manganese as impurities. Check analyses for steel or iron were available to the purchaser (but not necessarily required) for two lengths of pipe for each lot of 400 or less for sizes greater than 2-inch but less than 6-inch diameter and for two lengths for each lot of 200 or less for sizes of 6-inches and up. Analyses could also be done on the skelp instead of the pipe if agreeable to both the manufacturer and the purchaser. Reanalyses were permitted on two additional pieces if one of the first analyses did not pass. Ladle analyses for open-hearth steel were to be made available upon request. A comparable analysis (if one were made by a method not specified) was to be made available for Bessemer steel upon request.

The tensile properties which appeared in the first API 5L Specification are also shown in Table 9-1 At the time of the First Edition the specimens for all sizes of pipe were to be longitudinally oriented strips (unless the whole pipe was the specimen as was the case for small-diameter materials). The strips were to be removed at a location 90 degree from the weld for welded pipe. All tensile tests were tested "cold" (assumed to mean at ambient temperature rather than at some elevated temperature). The yield strength was determined by the "drop of the beam" method, by dividers, or by any other approved method. Elongation was to be measured on an 8-inch gage length for welded pipe and on a 2-inch gage length for seamless pipe. The tests for welded pipe could be made on the skelp by agreement between the purchaser and the manufacturer. Retests were permitted; two additional specimens from the same lot were required to meet the specification. Retests were permitted if the elongation was not met and the specimen failed outside of the middle third of the gage length. The required yield strength of Grade B (seamless) pipe at the time was 40,000 psi, not 35,000 psi, as it is today.

Hydrostatic testing of each pipe was required. The test pressures were fixed values shown in the various pipe tables and those values were limited by the allowable fiber stress ranges shown in Table 9-1.  In most cases the test pressures corresponded to hoop stress levels ranging from 40 percent of SMYS to 60 percent of SMYS.  The tests were to be maintained for not less than 5 seconds and lap-welded pipe was to be struck near both ends while under pressure with a 2-pound hammer.

The crop end of each length of lap-welded steel or open-hearth iron pipe was to be subjected to a flattening test. With the weld placed at the point of maximum bending, each end-crop ring was to be pressed diametrically between parallel plates and flattened to ⅔ of its original diameter with no opening in the weld.  Continued flattening to ⅓ of the original diameter without a crack or break elsewhere than the weld was required. Repeated tests were permitted until the requirements were met as long as 80 percent of the length after the first crop remained. A similar but somewhat less restrictive test was to be made on lap-welded pipe made from wrought-iron.

Butt-welded pipe in diameters over 2-inches was to be subjected to a ring flattening test somewhat similar to that for lap-welded pipe.  For smaller-diameter butt-welded pipe, a test involving bending a full-length piece around a 12-times-diameter mandrel to 90 degrees with no weld opening was required.  In both of these types of tests, the weld was to be placed at 45 degree to the maximum stress points.

The weights of pipe were to be no more than 3½ percent under the weights (per unit length) listed nor more than 10 percent over.  Lengths were not to be less than 9 feet for any individual lengths of plain end pipe, and the average length of plain-end pipe in an order was not to be less than 17 feet, 6 inches.  Lap-welded pipe and butt-welded pipe were generally made in lengths of 20 feet or less.  It was not until the advent of electric welded pipe and modern seamless pipe that 40 feet became the common length, and some of the early electric welded pipe was made in 30 foot lengths if it was formed by pyramid rolls. Apparently, the precedent was set however, to call a 20-foot piece a "random" length leading to the use of double-random lengths to mean, nominally, 40 feet and triple-random lengths to mean, nominally, 60 feet.

Workmanship standards in the First Edition of the API 5L specification required all skelp for lap-welded pipe to scarfed (i.e., the edges had to be beveled or tapered prior to pipe forming). Diameter variations were limited to 1 percent and the under tolerance for wall thickness was 12½ percent. The finished pipe was to be "reasonably straight and free from injurious defects". Defects mentioned included burnt material, bad welds, sand pits, ball cuts, pits, cinder spots, liquor marks, blister, slivers, and laminations. Seamless pipe was to be "free from injurious seams". Injurious was defined for sand pits, liquor marks, and cinder spots as being in excess of 12½ percent of the nominal wall thickness. Repairs to the body of the pipe by deposited weld metal were permitted for defects up to 33⅓ percent of the wall thickness.

In 1928, electric girth welds had not yet evolved as the common pipe-joining method. Therefore, the early API specifications and the First Edition in particular addressed threaded pipe in great detail. A thorough discussion of this material is not warranted herein, however, because very few major pipelines were ever constructed with threaded couplings. Plain-end pipe was generally made in these early times with 45 degree bevels rather than 30 degrees as is now common. The reason is that most pipelines in the 1920s and early 1930s were constructed with mechanical couplings and/or acetylene girth welds. Plain-end pipe to be used for such applications was required to pass ring-gage tests at the ends.

The sizes of pipe in the first specification (as now) were listed in two categories, threaded pipe and plain-end pipe.

Threaded pipe then (as now) came in a few standard sizes ranging in diameter from ⅛-inch (nominal) to 20-inch outside diameter. In the First Edition of the API specification, the sizes listed for "standard" plain-end pipe ranged from 3½-inch outside diameter to 20-inch outside diameter, but it was implied that larger sizes could be obtained.

Marking of each length of pipe was required. The information required was length, test pressure, and class of material (i.e., butt-welded, lap-welded, or seamless, and if seamless pipe, whether it was Grade A, B, or C). The A.P.I. monogram was also to be included as was the manufacturer's identifying mark. Marking of the manufacturer's name or mark, the API Monogram, and the grade (A, B, or C) if seamless was done by metal die stamps in ¼-inch high numbers one foot from the end of the joint. Paint stenciling was used to mark the length, test pressure, weight, and class of material. For bundled pipe the information could be provided on a metal tag.

Under the section titled "Inspection and Rejection", the rights of the purchaser were stated much as they remain today, namely, the granting of access to the manufacturer's facilities and giving of reasonable notification prior to the conducting of tests. The right of rejection based on the discovery of injurious defects or on failure to perform when properly applied in service was also stated in much the same form as it appears today.

The Second Edition of the specification was issued in January, 1929; and it was virtually unchanged from the First Edition. The required elongation for Bessemer steel was reduced from 20 percent to 18 percent for smaller butt-welded sizes.

## The Third Edition

In January, 1930, the Third Edition of the specification was issued. The most significant changes as-issued involved a carbon content limit and a tensile property change for Grade B pipe, the addition of "extra strong" (meaning thicker material) pipe, and the inclusion of a list of "companies authorized to use the A.P.I. Monogram on line pipe made in accordance with A.P.I. Standards No. 5-L".

A limit on carbon content of 0.30 percent by weight (ladle or check analysis) was placed on Grade B pipe in recognition of the increasing use of field welding to join pipe during construction. A footnote was also included to suggest that purchasers consider specifying a carbon limit on Grade C as well if it was to be field welded. The tensile requirements for Grade B were also changed. The minimum tensile strength was lowered from 70,000 to 65,000 psi, and the minimum yield strength was lowered from 40,000 to 38,000 psi probably because the higher prior limits may have been hard to meet once the carbon content was limited.

A "Supplement No. 1" to the Third Edition was issued with an issue date of February, 1931. Most of the items corrected by this supplement have no significance to this discussion. Some of the changes related to diameters dropped from tables and wall thicknesses added.

Others related to thread protectors and the dropping of some "working pressures" and "factors of safety". One change related to marking was important, however. First, it was stated that a manufacturer would forfeit the right to use the A.P.I. Monogram if it was not used on material manufactured or sold in accordance with the specifications within a 12 month period after the authority had been granted.

## The Fourth and Fifth Editions

When the Fourth Edition of A.P.I. Standard No. 5-L appeared in July, 1931, it was virtually the same as the Third Edition as modified by Supplement No. 1 to the Third Edition. However, significant changes appeared with Supplement No. 1 to the Fourth Edition (dated July, 1932). Check analysis frequency was increased to two lengths for each lot of 100 for larger pipe sizes. The required elongation of Grade A seamless pipe was lowered to 30 percent from 40 percent. Tensile test width was set at 1½ inches. The number of tensile tests was increased to one for each lot of 100 pieces for larger pipe sizes. All references to "working pressures" were deleted. Open-hearth iron was added to the list of materials permitted for the manufacture of seamless pipe. Such pipe was to have the same tensile properties and hydrostatic test pressures as welded pipe made from open hearth iron.

The most important changes, however, were the addition of "electric welded pipe", and the reduction of the required tensile properties for Grade B pipe. The process of electric welding was added without any definition as to the process with a requirement that it be made of open-hearth steel unless otherwise specified. It could be made as Grade A, Grade B or Grade C. The word "furnace" was inserted in front of butt-welded and lap-welded to distinguish those processes from electric welded. With respect to Grade B, the limits on manganese were changed from 0.30 minimum, 1.50 maximum to 0.30 minimum, 0.70 maximum for electric welded pipe only (the old limits still applied to seamless). The limit on phosphorous for seamless and electric welded Grade B was changed from 0.04 to 0.045. The tensile properties for Grade B were changed as follows; tensile strength, 65,000 to 60,000 psi; yield strength, 38,000 to

35,000 psi. These tensile properties have remained the same ever since for Grade B material.

The hydrostatic test hoop stress levels were lowered. The average range was 18,000 to 16,000 psi and a provision was added to the effect that no mill hydrostatic test pressure could exceed 80 percent of SMYS. The provision for hitting the weld with a two-pound hammer at each end was extended to electric welded pipe.

Flattening tests were extended to electric welded pipe, but the cut off ring was specified to be a minimum of 4 inches (no limit had existed for lap-welded or butt-welded pipe). Also, testing of one length in each lot of 200 was specified for electric welded pipe instead of one from each end of each length as with butt-welded or lap-welded pipe. Presumably this was done because electric welding was expected to produce more uniform welds than the furnace weld processes.

A weld tensile test was added to evaluated electric welded seams. The tensile test was to be taken at right angles to the weld, with the weld located in the center of the specimen. The ultimate tensile strength of the specimen was to equal or exceed the minimum specified tensile strength of the pipe.

Under workmanship, a limit on length of 25 percent of diameter was placed on defects in seamless pipe for which chipping and repair by deposited weld metal was permitted. Electric welded pipe was included in the list for which injurious defects would not be accepted. However, repair of electric weld seam defects was permitted even for leaks if the purchaser gave approval.

The A.P.I. Standard No. 5-L Fifth Edition of the A.P.I. Line Pipe Specification was dated January, 1934. It was essentially the same as the Fourth Edition as supplemented in July, 1932. One change worth noting was that plain-end pipe would henceforth have 37½ degree bevels instead of 45 degree bevels. This was undoubtedly done in deference to the increasing use of electric girth welding. Another change was in the area of marking. The various classes of pipe were to be marked as follows:

L - Furnace Welded
E - Electric Welded
S - Seamless
I - Wrought Iron or Open-Hearth Iron.

A third change was that the statement "All tests shall be done cold" was changed to "All physical tests shall be made at room temperature".

**The Sixth Edition**

The Sixth Edition was dated August, 1935 and was initially about the same as the Fifth Edition. It added a definition for wrought iron which was here-to-fore not included. The definition was: wrought iron is defined as a ferrous material, aggregated from a solidifying mass of pasty particles of highly refined metallic iron with which, without subsequent fusion, is incorporated a minutely and uniformly distributed quantity of slag (A.S.T.M. A-81). It also added a section called "Adjustments under Inspection and Rejection". This section put the burden of assuring compliance on the purchaser. It also provided a complaint procedure involving review and arbitration by groups within A.P.I.

A number of supplements to the Sixth Edition were produced. Only two of these were located, Supplement No. 3 (September, 1938) and Supplement No. 6 (September, 1939). The impact of the missing supplements can be assessed by examining

the Seventh Edition.  However, some of the changes in the two
supplements obtained are worth mentioning.  Supplement No. 3
permitted the marking of pipe with metric units at the option of
the manufacturer or purchaser.  It also enhanced the purchaser's
rights of inspection and rejection.  It did so by permitting
tests at locations other than at the manufacturer's facility.
The purchaser was allowed to test or inspect the pipe at any time
and place and reject it if it failed to meet the requirements.
Supplement No. 6 changed the definition of yield strength from
the "drop of beam method" to 0.5 percent total strain, the
current definition.

**The Seventh Edition**

With the issuance of the Seventh Edition dated April
1940 the periods between A.P.I. were dropped.  The title of this
edition was API Standard No. 5-L, Seventh Edition, API Line Pipe
Specification.  The separate chemistry for electric welded Grade
B was dropped and electric welded pipe was once again grouped
with seamless Grade B with respect to chemical requirements.
Tables with A.S.A. (American Standards Association) pipe
thickness schedules were included as Appendix A for information
only.

In Supplement No. 2 to the Seventh Edition (the only
supplement we were able to find), the standard plain-end bevel
was changed from 37½ to 30 degrees where it remains today.  The
phosphorus content of Grades A and C was changed from 0.04 to
0.045 to agree with Grade B.  Under Appendix D, a section called
"Use of Monogram" was added which stated conditions for
cancellation of a license.

**The Eighth Edition**

The Eighth Edition was dated May, 1942. For the first
time, pipe processes were listed in a table. All minimum
hydrostatic test pressures became 60 percent of SMYS with the
provision that higher test pressures could be specified up to
80 percent of SMYS but in no case more than 2500 psi. Tolerances
on thickness, diameter, lengths and other properties were put
into tables.

Almost immediately, changes to the Eighth Edition were
proposed entitled "War Emergency Measures". For the most part
these were aimed at allowing the various products to be made by a
wider variety of processes. The additions to the table of
allowed materials and processes were as follows.

| Material | 8th Edition as-written | Processes Allowed | After War Emergency Measures the following were added |
|----------|------------------------|-------------------|-------------------------------------------------------|
| Steel | Bessemer (B) | F,L | E,S |
| | Electric (E) | S,F,L | E |
| | Open Hearth* | E,S,F,L | |
| Iron | Open Hearth (I) | S,F,L | E |
| | Wrought Iron (WI) | F,L | |

F = Butt-welded, L= Lap-welded, S = Seamless, E = Electric
welded

\* No symbol was used for open-hearth steel.

The War Emergency Measures were instituted in
Supplement No. 1 (November, 1942) which also contained other
changes. The minimum elongations for Grades A and B were made
dependent on wall thickness and put in a table. Small size plain-

end pipe (<3-inch diameter) was permitted. The tables for information on ASA standard sizes were deleted (thus they appeared only once, in the Eighth Edition). Lastly, the standards for marking were stated more clearly and in more detail. The manufacturer's mark, the API Monogram, the grade, the class, and the material (except for open-hearth steel) were to be die-stamped. The size, weight per foot, grade, class, material, and length were to be paint-stenciled.

## The Ninth Edition

In the Ninth Edition (dated August, 1944), the following definition was given for the first time for electric welded pipe "Electric welded pipe is defined as pipe having longitudinal seams joined by electric welding without the addition of extraneous material, such as flash welding (first mention of flash-welding) or continuous electric resistance welding". A requirement was added that Bessemer steel was to be killed deoxidized acid Bessemer steel. A provision was made to allow (but not to require) mill hydrostatic tests to be longer than 5 seconds. Finally, for the first time, the transverse tensile test, of the pipe material appeared. In addition to the longitudinal tensile test, a transverse tensile test was to be made on electric-welded pipe 8⅝ inches and larger.

## The Tenth Edition

The Tenth Edition was dated August, 1945. It permitted butt-welded pipe in 3½ and 4-inch diameters. It introduced a non-mandatory flattening test of a ring taken from the middle of electric-welded pipe, and it allowed the purchaser to refuse to take jointers of plain-end pipe. Also, equations were provided for the first time for calculating elongations of Grade A and Grade B materials.

In a supplement dated July 1946, the processes allowed under War Emergency Measurers were made permanent. That is, manufacturer's could continue to use Bessemer steel for seamless, and electric welded pipe and electric steel and open-hearth iron for electric-welded pipe. The burden for compliance with the specifications was shifted from the purchaser to the manufacturer, however, it was still left to the purchaser to make appropriate inspections and evaluations by means of choice. In return the purchaser's right of rejection was clarified. Any material which did not comply with the specification as determined by the purchaser's inspection could be rejected. In a second supplement (dated November 1946), the existence of a new specification "5-LX High-Test Line Pipe (Tentative)" was alluded to.

The Tenth Edition was the last of the "pocket-sized" versions of the API Specifications for line pipe. When the next edition was issued in 1949 it was 8 x 10½ inches, and it had been completely reorganized.

## 5L Specifications in the Period from 1949 Through 1982

### Eleventh Through Fourteenth Editions

When the Eleventh Edition of the 5L specification appeared in May 1949, it was arranged in an 8-inch x 10½-inch format with tables and text looking more like those in the 1995 5L specification than those in the Tenth Edition. The substance of the requirements did not change very much, however, in the period between the Tenth and the Fourteenth Edition. One addition to the Eleventh Edition was a footnote to the table on plain-end pipe sizes which indicated that API lap-welded pipe was no longer made in diameters exceeding 16 inches. Another change in the Eleventh Edition was the introduction of the transverse flattened tensile specimen for measuring the yield and ultimate strengths of electric-welded pipe of sizes 8⅝-inch outside diameter and up. Heretofore, longitudinal tensile tests had been specified for yield and ultimate strengths (the transverse weld-tensile test had been introduced earlier in the Fourth Edition). The longitudinal tensile test was still to be used for all seamless, butt-welded, and lap-welded pipe and for electric welded pipe in sizes smaller than 8⅝ inch. Finally, a table of "double extra strong" pipe sizes was added.

The only significant change in the Twelfth Edition (March, 1951) was the provision for permitting the ring-expansion test as an alternate means of determining the yield strength of pipe. In the Thirteenth Edition (March, 1954), continuous-welded pipe is recognized for the first time as a variation of butt-welded pipe.

The Fourteenth Edition, issued in March of 1955, contained several significant changes from the Eleventh through Thirteenth Editions. Grade C pipe (45,000 psi yield strength) was deleted no doubt because by this time the 5LX specification for high strength grades was being fully utilized. The chemical limits for ladle analyses were lowered for most products especially with respect to carbon contents (compare 12th and 14th editions in Table 9-1). Plus tolerances for chemical check analyses were provided for the first time, and the first mention of cold expansion appears in the flattening test paragraphs. Other changes included the addition of the limit of a ¼-inch-deep dent of ½-diameter length as an injurious defect, a change in the minimum wall thickness tolerance on welded pipe of 20-inch outside diameter and up from -12.5 to -10 percent, a change in the mill hydrotest hold time from 5 to 10 seconds for pipe of

20-inch outside diameter and up, and the requirement that die
stamping markings be done with a blunt stamp with rounded edges.
The Fourteenth Edition appeared to reflect a change in philosophy
geared toward achieving a higher quality product.  Over the next
25 years many more such changes occurred as detailed below.

**Fifteenth Through Eighteenth Editions**

The first mention of a limit on yield-to-ultimate
strength ratio (0.85) for cold expanded pipe appeared in the
Fifteenth Edition (March 1956).  Plus tolerances for wall
thickness were also added in that edition (+15 percent,
-12.5 percent for all seamless pipe and welded pipe of less than
20-inch outside diameter; +15 percent, -10 percent for 20-inch
outside diameter and up).  A new table appeared in the Sixteenth
Edition (April, 1957) giving length tolerances and providing, for
the first time, limits for double random "lengths, a recognition
of the continuing shift from the 20-foot lengths typically
produced by the older furnace welding processes to the more
common 40-foot lengths typically produced by electric-welded and
seamless pipe producers.  Recognized for the first time in the
Seventeenth Edition were the flash-welding process for making
electric-welded pipe and, on a tentative basis, the basic oxygen
method for steel making.  The 5L requirements were also
tentatively extended to electric-induction-welded pipe.
Manufacturers were not yet allowed to "Monogram" a material made
with basic oxygen steel or an electric-induction-welded material.
In a supplement to the Seventeenth Edition dated March, 1959, the
limit on manganese for certain materials was reduced to
1.15 percent from 1.35 percent and a limit of 1 percent out-of-
roundness was placed on materials of 20-inch outside diameter and
up.

Between February, 1960 when the Eighteenth Edition was published and January, 1961 when a supplement to the Eighteenth Edition was issued, the following changes were made. Basic oxygen steel making was fully accepted. The flattening test procedure for nonexpanded flash-welded and single length electric-resistance-welded materials was modified such that the test on one end of the single piece involved placing the weld at 90 degree to the press surfaces and placing the weld at 0 degree to the press surfaces for the specimen taken from the other end of the piece. For nonexpanded electric-resistance welded materials made continuously from coiled skelp and cut to length, two intermediate tests were added to the two previously required, one for each end of the coil. Open hearth iron was dropped from the specification and the use of wrought iron was restricted to butt-welded and lap-welded materials. A definition was provided for the first time for seamless pipe: "a wrought steel tubular product made without a welded seam. It is manufactured by hot working steel or, if necessary, by subsequent cold finishing the hot-worked tubular product to produce the desired shape, dimensions and properties." Electric welded pipe was redefined to include electric-flash welding, electric-resistance welding, or electric induction welding. As before, the definition referred to a material with "one longitudinal seam" made "without the addition of extraneous metal". Lap-welding and butt-welding were defined as "pipe with one longitudinal seam formed by mechanical pressure to make the welded junction, the edges being furnace heated to the welding temperature prior to welding". It is noted that this definition would allow hammer-welded pipe although the latter never appeared in any API 5L specification. The induction-welding process remained tentative and was further limited to sizes 6⅝-inch outside diameter and smaller with wall thicknesses of 0.280-inch or less and applicable to Grade B only.

## Nineteenth Edition

This edition appear in March, 1962.  The changes
reflect the continuing evolution of the pipeline industry.  The
requirements on threads and gaging practice no longer appeared in
the 5L standard after the Eighteenth Edition.  They were moved to
a separate standard, API Std. 5B. Submerged-arc welded line pipe
(one outside diameter and one inside diameter pass) was added and
the cold expansion process was mentioned explicitly in a separate
paragraph in the Nineteenth Edition.  A longitudinal weld tensile
and a guided bend test were added as a required test for SAW-seam
pipe.  Lap-welded pipe made from steel was dropped leaving that
made from wrought iron as the only approved lap-welded material.
The pipe size tables were enlarged to include sizes up to 36-inch
outside diameter.  The major sections were renumbered with Arabic
numerals instead of Roman Numerals, and the paragraphs were
numbered separately within each section rather than consecutively
as they had been through the Eighteenth Edition.

The most significant changes in the Nineteenth Edition
occurred in the section on hydrostatic tests.  A "verification"
paragraph was added which required a permanent recording of the
test pressure of each piece and an interlocking system that would
identify a piece as tested if and only if is was indeed tested.
The records were to be made available to the purchaser on
request.  Finally, an alternate test pressure equivalent to
75 percent of SMYS was added to the standard test pressure
corresponding to 60 percent of SMYS.

## Twentieth Edition

Changes observed between the Nineteenth Edition and the
Twentieth Edition (March, 1963) were along the lines of previous

evolutionary changes. Both lap-welded pipe and wrought iron were deleted. Drawings were added to illustrate tensile test specimens. An appendix with pipe tables in metric dimensions was added. Under tensile tests reference was made, for the first time, to ASTM A370, Mechanical Testing of Steel Products, Supplement II, Steel Tubular Products as the standard for conducting such tests.

The other set of changes was revolutionary in nature. In place of the previous "workmanship" requirements which specified no explicit methods for detecting injurious defects, an explicit menu of nondestructive inspection methods was set forth. Radiographic inspection was specified for the longitudinal seam of SAW pipe. Ultrasonic, eddy-current, or magnetic particle inspection was specified for the longitudinal seam of any electric-welded material. Acceptance or rejection of the product was based upon specific size limits for certain types of flaws. The flaw types specifically mentioned included undercut, misaligned weld and outside bead height for SAW pipe; and offset edges and incomplete fusion for all seam types. Any defect having a depth of more than 12½ percent was still considered injurious, and all cracks, lack of penetration and lack of fusion were rejectable. The previous limits on dents were retained.

## Twenty-First Through Twenty-Third Editions

The changes in the contents of these three editions were mostly of the evolutionary type. In the Twenty-First Edition (March, 1965), electric-induction welded pipe was accepted in sizes not exceeding 6⅝-inch outside diameter and wall thicknesses not exceeding 0.280-inch for Grade B material only. Also, this edition was the first to allude to the existence of a 5LS (spiral-welded pipe) specification and RP 5L1, the recommended practice for rail car loading of pipe.

In the Twenty-Second Edition (March, 1967), the nondestructive testing requirements which first appeared in the Twentieth Edition were set apart in a separate section (Section 9). Workmanship, Visual Inspection and Repair became Section 10. Additionally, limitations were imposed on ID bead height for SAW seams and ID trim for ERW seams. This edition also contains the first mention of a prohibition of hot flattening for transverse tensile tests and the first mention of a requirement for postweld heat treatment of ERW seams for Grade B material. The latter required heating to 1000°F minimum to assure the absence of untempered martensite. It did not, however, require "normalizing" of the seam.

In the Twenty-Third Edition (March, 1968) a provision was added to require the tensile test strain rate to conform to ASTM A370. The minimum hydrostatic test pressure was raised from 2500 to 2800 psi. Drawings of the reference standards for ultrasonic and electromagnetic inspection were added to Section 9. Elongation tables were placed in an appendix, and the plus tolerance on wall thickness for 20-inch outside diameter pipe and up was raised from +15 to +17.5 percent.

**Twenty-Fourth Edition**

The Twenty-Fourth Edition (April, 1969) contained a substantial number of changes. A supplemental requirement SR-4 was added with nonmandatory requirements for nondestructive inspection of seamless pipe. The guided bend jig drawing was put into an appendix. Electric-induction welded pipe was accepted without size or grade restrictions. The size tables were expanded to include diameters up to and including 44-inch outside diameter. The separate tables for extra-strong pipe and double extra strong pipe were eliminated, and these sizes were incorporated into the main tables with specific notations. Cold die stamping was prohibited on nonheat-treated Grade A and Grade B materials and on all pipe materials with wall thicknesses of less than 0.156-inch.

The most significant changes in the Twenty-Fourth Edition were associated with changes in the types of steel permitted for the manufacture of line pipe. For one thing, Bessemer Steel was deleted for the manufacture of butt-welded pipe and open-hearth Class I and Class II materials were eliminated altogether. In their place, a new grade, Grade A25, was added for which the minimum yield strength was 25,000 psi and the minimum ultimate strength was 45,000 psi. The carbon content of this material was more restricted than those of the deleted materials. Grade A25 material was not permitted to be used to manufacture SAW pipe, but the manufacture of butt-welded (and continuous-welded) pipe was restricted to the new Grade A25 material. In addition heat treatment was now addressed as a processing variable and the heat treatments allowed were as-rolled, normalized, normalized and tempered, subcritical stress relief, and subcritical age hardening.

Finally, it is noted that two new publications were referenced in the "other API specifications list": Bulletin 5T1 on Nondestructive Testing Terminology and RP5L3, Recommended Practice for Conducting Drop-Weight Tear Test on Line Pipe.

## Twenty-Fifth Through Thirty-Second Editions

During the next 12 years the changes were relatively minor and not of much consequence to pipe quality. Briefly they were as follows.

### Twenty-Fifth Edition (April, 1970)

- Bessemer Steel eliminated entirely leaving basic oxygen, open hearth, and electric furnace steel making as the only approved methods.

- The pipe size table was expanded to include 48-inch outside diameter pipe.

- The procedure for repair welding the seam of an ERW pipe was specified in detail whereas previously it had been by agreement between the purchaser and the manufacturer.

### Twenty-Sixth Edition (April, 1971)

- A hard spot 2-inch in diameter or larger was defined as an injurious defect for pipes of 20-inch outside diameter and up.

### Twenty-Seventh Edition (March, 1973)

- Double seam SAW pipe greater than 36-inch outside diameter was included.

- The weld ductility test was added for ERW and flash-welded pipe.

- A specification for belled ends was added.

- Specification 5LU for Ultra High Test Heat-Treated Line Pipe was added to the list of other API specifications.

### Supplement to the Twenty-Seventh Edition (March, 1974)

- A arc burn was declared to be an injurious defect but contact burns were not.

- A procedure for qualifying fluoroscopic operators was added.

**Twenty-Eighth Edition (March, 1975)**

- A fluoroscopy evaluation procedure was added.

**Supplement to Twenty-Eighth Edition (March, 1976)**

- Pipe size table expanded to include 64-inch outside diameter pipe.

**Twenty-Ninth Edition (March, 1977)**

- The term injurious defect (rejectable) was replaced by two terms: defect (rejectable) and imperfection (not rejectable).

**Thirtieth Edition (March, 1978)**

- Gas-Metal-arc welding was accepted as a seam welding procedure for Grades A and B only.

- The term "heat" analysis replaced ladle analysis.

- The term "product" analysis replaced check analysis.

- Acceptable chemical analysis procedures were added with the provision that each was traceable to standards ASTM E-59 and ASTM E-350 for calibration.

**Thirty-First Edition (March, 1980)**

- Supplement dated March, 1981 added the transverse round bar tensile test removed without flattening to the acceptable means of measuring yield and ultimate strengths.

**Thirty-Second Edition (March, 1982)**

- This was the last edition of Specification 5L which covered only Grades A25, A and B. When the Thirty-Third Edition appeared in March of

1983 it addressed all types and grades of line pipe. The separate documents for 5LX and 5LS were discontinued.

## 5LX Specifications in the Period from 1948 through 1982

**The First Edition**

In February 1948 more than 2 years after the Tenth Edition of API 5L was issued and about a year before the Eleventh Edition of API 5L appeared, the first Tentative Standard 5LX, Specification for High Test Line Pipe was issued. This document was issued in the same 8 x 10½ format as the Eleventh Edition of Specification 5L. Unlike the 5L document which had 34 pages, the first 5LX tentative standard contained only 11 pages. Since it addressed only one specific grade (X42) and applied only to plain end pipe (beveled for welding), the tables in it were simple and few in number. It alluded to grades higher than X42 but these were to be made strictly by and according to agreements between the manufacture and the purchaser.

The manufacturing processes included seamless pipe or mill-welded pipe comprised of cold-preformed skelp, longitudinally welded by electric-flash welding, continuous electric-resistance welding, and submerged-arc welding. The types of steel included open-hearth, electric furnace, and killed deoxidized acid Bessemer steel.

The chemical property limits for carbon, manganese, phosphorus and sulfur (ladle and check analyses) were given in a table. They are repeated herein in Table 9-2. The limits on carbon and manganese have changed little over the years. For example, the upper limit on carbon then was 0.33 percent by weight (check analysis) versus 0.32 for welded, cold expanded X42, X46, and X52 in 1995 (40th edition). Similarly, the upper limit on manganese was 1.28 percent then versus 1.35 for welded, cold-expanded X42, X46 and X52 in 1995. The phosphorus and sulfur limits then (0.115 and 0.065 percent respectively) have been lowered considerably over the years, however. Both are currently limited to 0.04 percent. The changes in the limits on these nonmetallic elements reflect the concern which developed over the years regarding the adverse effects of "dirty" steels on the performance of the material in service. The sampling rates were set at one of each of two lengths from each lot of 200 lengths for small diameter pipe (6⅝ to 12¾) and one of each of two lengths from each lot of 100 lengths for pipe of 14-inch diameter and larger.

The tensile properties for Grade X42 were established as 42,000 psi for its minimum yield strength and 60,000 psi for its minimum tensile strength, the same as they are today. The sampling rate was the same as that for chemical analysis. Yield strength was to be determined as the stress required to produce 0.5 percent total strain. Longitudinal specimens were prescribed for all hot-rolled seamless products and for welded and cold-expanded seamless pipe of 6⅝-inch diameter. Transverse specimens were prescribed for welded pipe and cold-expanded seamless pipe of 8⅝-inch diameter and larger. Ring tensile specimens were permitted for the determination of yield strengths. A transverse tensile test of the weld (same sampling rate as the pipe body test) was required for tensile strength only. All tests were to be conducted at room temperature.

Flattening tests were prescribed for the crop ends cut from each end of each length. The weld was to be placed at the point of maximum bending. No opening in the weld was permitted until the ring was flattened to ⅔ of the original outside diameter. No crack or breaks were permitted in the metal elsewhere than the weld until the ring was flattened to ⅓ of the original outside diameter. The ring was to be pressed completely flat with no evidence of laminations or burnt material. As an alternative to testing the crop ends, by agreement between the manufacturer and the purchaser, flattening tests could be made on one 4-inch wide ring (or wider) cut from the end of one length per each lot of 200 lengths or from the middle of the number of lengths agreed upon.

Each length of pipe was to be subjected to a 5 second hydrostatic test to 85 percent of SMYS (not to exceed 3000 psi). For welded pipe the weld was to be stuck with a 2-pound hammer near the weld at both ends during the test.

A table was given for dimensions, weights, and hydrostatic test pressures. In the first edition the sizes listed included 6⅝, 8⅝, 10¾, 12¾, 14, 16, 18, 20, 22, 24, and 26-inch outside diameter. The diameter tolerance was ±1 percent for all sizes. The wall thickness tolerance was -12.5 percent for all sizes. Lengths were to be no shorter than 9 feet and the average of the lot was to be not less than 17 feet, 6 inches.

Ends were to be furnished beveled to 30 degrees from a line perpendicular to the axis of the pipe (+5, -0 degrees) with a 1/16-inch land (±1/32 inch).

Welded pipe was to be reasonably straight and free from injurious defects. Any defect was considered injurious if its depth was greater than 12½ percent of the wall thickness. Repairs to both pipe body defects and weld defects were permitted within certain limitations.

The marking requirements called for die stamping and paint stenciling of certain information on each length of pipe. The information to be die stamped included the manufacturer's mark, the grade, the process of manufacturer (seamless or electric welded), and the type of steel (0 for open hearth or electric furnace steel, B for Bessemer steel). The die stamping was to be placed not less than 8 nor more than 12 inches from the end of the pipe and the markings were to be ¼ inch in height.

The information to be paint stenciled included outside diameter, weight per foot, grade, process of manufacture, type of steel, length and hydrostatic test pressure (the latter to be included only if it exceeded the normal requirement). As the first edition was a "tentative" specification the API Monogram was not permitted.

Concerning inspection, the purchaser was given the right to free entry to the manufacturer's facilities at which his pipe was being made. Reasonable notice of the time of the run was to be given. Reasonable facilities were to be provided free of charge for the purchaser to satisfy himself that the pipe was being made according to the specification. The purchaser had the right to reject material with injurious defects and to receive compensation for any material which proved to be defective when properly applied in service. The manufacture was responsible for compliance with the specification. The purchaser could make any investigation necessary to satisfy himself of compliance and could reject any material which did not comply with the specification.

Except for an appendix dealing with jointers that was all there was to the First Edition.

**The Second Edition**

The Second Edition of API STD 5LX was adopted as of May, 1949. This was a "standard" and was no longer tentative. While the First Edition alluded to submerged-arc welded pipe, the Second Edition provided the following detailed definition. "Submerged-arc welded pipe is defined as pipe welded by the submerged-arc welding process with *at least two weld passes*, of which one is on the outside of the pipe and one on the inside. The submerged-arc welding *shall be started and stopped on plates* tack-welded to the ends of the pipe (where welding is continuous from one length to another such plates need be used only at the pipe ends where the welding is started or stopped). All cracks or other defects in submerged-arc welds shall be completely removed and repaired by either manual arc-welding using coated electrodes or by submerged-arc welding". The italicized phrases were not italicized in the specification. They are emphasized herein because this implied the nonacceptance of single-submerged-arc welded pipe and the nonacceptance of "squirt" welds to finish the ends of pipes.

The chemical analysis table was expanded to include more restrictive limits on phosphorus for open-hearth and electric-furnace steels. For a check analysis the limit was lowered from 0.115 to 0.055. The limit for rephosphorized open-hearth steel was set at 0.090 and the limit for Bessemer steel was set at 0.110. The tensile requirements table was unchanged. The chemistry and the tensile requirements were still for Grade X42 only. The chemistry and tensile requirements for higher grades were to conform to the requirements agreed upon between the purchaser and the manufacturer.

Requirements for the flattening tests were changed slightly. Whereas the First Edition specified the test for all welded pipe, the Second Edition specifically exempted submerged-arc welded pipe by referring only to electric-flash welded pipe and electric-resistance welded (ERW) pipe.

A weld bend test was added for submerged-arc welded pipe. A root-bend and face-bend test were to be made according to the ASA B31.1 Code. One of each was to be made from each lot of 50 lengths.

The remaining changes to the Second Edition related to API monogram. The monogram was required now that the specification was no longer tentative. An Appendix B was added in anticipation of a list of authorized manufacturers. No manufacturers were listed as yet, however, because the standard had just been approved. Other materials were added concerning the use of the monogram and a statement of qualifications for manufacturers to use in applying for the right to use the monogram.

A supplement to the Second Edition was issued in December, 1949. It changed the markings for type of steel. E was now to be used for electric-furnace steel and no marking was to be used for open hearth steel whereas O had been used previously for both. The supplement also contained an application form for permission to use the monogram, a form of certificate of authority to use the monogram and a list of causes for cancellation of monogram rights.

**Third Edition**

The Third Edition of the 5LX Specification was issued in March, 1951. It differed from the Second Edition in the following respects. Specific guidance was given for conducting a guided bend test on the weld for submerged-arc welded pipe. The specimen dimensions and bending guide were illustrated in new figures. The bends were to be accomplished with no cracks larger than ⅛ inch allowed. Corner cracks were excepted. The pipe table was expanded to include 28 through 36-inch outside diameter pipe, and the first list of 5LX-certified manufacturers appeared.

In a supplement to the Third Edition issued in January, 1952, an interesting change was instituted. Previously, under the "Process of Manufacturer", the definition of sub-merged arc welding included language which required run-off tabs to start and end the welding and which required the complete removal and repair of all cracks and other defects in the seam. The supplement deleted these two provisions. This was probably done because at least two manufacturers at the time did not use run-off tabs and every weld they completed by "squirt" welding.

The workmanship requirements regarding the repair of defects were made more specific, the check analysis limits were changed for carbon from 0.33 to 0.34 and for manganese from 1.28 to 1.30. The weld bend test was changed from being mandatory to being by agreement, and the marking requirements where changed to permit inside diameter marking by agreement instead of or in addition to outside diameter marking.

## Fourth Edition

In the Fourth Edition, March, 1953, killed deoxidized basic Bessemer steel was allowed tentatively, and tensile requirements for Grade X46 and X52 were set. The weld bend specimen requirement regarding edge cracks was changed to limit them to ¼ inch whereas before edge cracks (i.e., corner cracks) were not considered. A nick-break requirement (by agreement), was added for submerged-arc welded seams, but this only appeared in this edition; it was subsequently dropped.

The mill hydrostatic test was altered such that materials with 8⅝-inch diameter or less were only required to be tested to 75 percent of SMYS as opposed to 85 percent of SMYS. Larger sizes were still required to be tested to 85 percent of SMYS. Marking requirements were altered to permit hot rolling in addition to die stamping and paint stencilling, and paint stencilling only for sizes larger than 12.75-inch was allowed.

A supplement to the Fourth Edition was issued in February, 1954. It contained minor changes aimed at cold expanded pipe including tightening the chemical limits and introducing a minimum yield to tensile strength ratio of 0.85. Also, the end bevel angle was changed from 30 degrees to 37½ degrees.

## Fifth Edition

In November, 1954, the Fifth Edition of API STD 5LX was issued. With this edition the chemistry table was expanded to include Grades X46 and X52 and chemistries for the various types of steel were specified.

The flattening test was altered such that the tests were to be done on the crop ends for nonexpanded single length flash-welded and ERW pipe and on both ends of the cut lengths of each lot of 20 lengths of nonexpanded ERW made continuously and cut into multiple lengths. For cold-expanded flash-welded and ERW pipe, the number of flattening tests remain the same, namely, one per length of each lot of 200 lengths.

A longitudinal weld tensile test was added for submerged-arc welded pipe. The specimen was shown in a new figure; it contained the weld oriented along its axis. To pass the test the specimen had to exhibit the same elongation as that required for the pipe body.

The hold time in the mill hydrostatic test was increased to 10 seconds (from 5 seconds) for materials of 20-inch outside diameter and up, and the wall thickness tolerances were altered. While -12.5 percent was retained for pipes of 18-inch outside diameter or less, a new tolerance of -10 percent was added for pipes of 20-inch outside diameter and up.

**Sixth Through Tenth Editions**

The next five editions and their supplements produced only minor changes, the most obvious of which are as follows.

**Sixth Edition (February, 1956)**

- Killed deoxidized basic Bessemer steel became acceptable.

- The minimum tensile strength for Grade X52 was set at 72,000 for diameters of 20 inches and up and for wall thicknesses of 0.375 inch or less.

- The mill hydrostatic test pressure was increased to 90 percent of SMYS for welded pipe 20 inches or more in diameter.

- The average length of 35 feet was set for double random lengths.

- A plus tolerance of 15 percent was added for wall thickness.

- The bevel angle was changed back to 30 degrees.

- A requirement was added that end welds for DSAW pipe lengths must be made by an automatic submerged-arc process or else be considered a repair weld.

**Seventh Edition (April, 1957).** The only changes were to include high-strength low allow steel made by the open hearth or electric furnace methods as an acceptable material, and to require two flattening tests, one with the weld at 90 degrees and one with the weld at 0 degree, per length for each cut length of continuously made multiple length nonexpanded ERW pipe.

**Eighth Edition (March, 1958).** Basic-oxygen steel was included as a tentative pipe material (API monogram not permitted). The following symbols were added under "Marking": M for high-strength low alloy steel and BO for basic oxygen steel.

**Supplement to Eighth Edition (March, 1959).** Seamless pipe of sizes 20-inch and larger was henceforth to be hydrostatically tested to 90 percent of SMYS.

**Ninth Edition (February, 1960).** The flattening test requirements for nonexpanded continuously made ERW pipe were changed again. In this edition the requirement became one per crop end with the weld at 90 degrees and two from intermediate rings with the weld at 0: This requirement is the same as that in the present (1995) standard.

**Supplement to Ninth Edition (January, 1961).** A definition was added for seamless pipe. "Seamless pipe is defined as a wrought steel tubular product made without a welded seam. It is manufactured by hot working steel or, if necessary, by subsequent cold finishing the hot worked tubular product to produce the desired shape, dimensions and properties".

**Tenth Edition (March, 1962).** Basic oxygen steel was accepted. The sections of the document were numbered with Arabic instead of Roman numerals, and paragraphs were numbered by section rather than sequentially. A section on hydrostatic test verification was added which required the test pressure and duration to be recorded by an interlocking system wherein the recording could take place only after the test had been completed in compliance with the requirements. The permanent record was to be made available for inspection by the purchaser's representative. Pipe sizes of 4½-inch outside diameter and 38, 40, and 42-inch outside diameters were added. The standard test

pressure for 4½-inch outside diameter pipe was to be 60 percent
of SMYS with an alternate test pressure of 75 percent of SMYS.

**Supplement to the Tenth Edition (July, 1962).** Carbon
steels with manganese contents of 1.35 percent or less (ladle
analysis) were included.

## Eleventh Edition

The Eleven Edition appeared in March, 1963.  It
contained three minor and one very major change from the Tenth
Edition.  Two of the minor changes involved the addition of a
figure to describe the tensile test specimens and the provision
for the tests to be conducted on specimens conforming to ASTM
A370: Mechanical Testing of Steel products, Supplement II, Steel
Tubular Products.  The third change involved the adding of an
appendix with a table of dimensions, weights, and test pressures
in metric units.  The major change related to the inclusion of
nondestructive testing requirements.  These were added to the
section on Inspection and Rejection and replaced the short
paragraphs on workmanship which appeared in prior editions.

Under the heading "General" the new Section 9 contained
the prior familiar paragraphs on giving notice, plant access,
rejection, and compliance.  These remained virtually the same as
before.  Two comprehensive new subsections called "Nondestructive
Inspection" and "Visual Inspection" were added.  Finally, the
last subsection, titled "Repair of Injurious Defects" contained
the same material as the pervious "repair of defects" paragraph
in prior editions.

Under "Nondestructive Inspection", 100 percent
inspection was mandated for the welded seams of DSAW, ERW and
flash-welded pipe.  The inspection was to be by radiological

methods for DSAW pipe and by ultrasonic, eddy-current, or magnetic particle methods for ERW and flash-welded pipe. For cold expanded pipe the inspection of the ends of the welds had to be repeated after expansion if the entire seam was inspected prior to cold expansion. Detailed descriptions of the equipment, the reference standards, and the acceptance limits for defects were given for all of the methods.

Under "Visual Inspection", various kinds and sizes of imperfections which constituted injurious (rejectable but possibly repairable) defects were listed. Any imperfection over 12½ percent of the wall thickness was considered injurious. All cracks and leaks were considered injurious. Size limits were given for dents, undercuts, offset plate edges, misalignment of the seams in DSAW pipe (off-seam weld) and height of the outside diameter weld bead.

## Supplement to the Eleventh Edition

The supplement to the Eleventh Edition issued March, 1964, provided for some changes in tensile test requirements and nondestructive testing requirements. The number of tensile tests for 4½-inch pipe was changed to 1 per lot of 400 (it was, formerly lumped with 6⅝-inch pipe for which 1 per lot of 200 was required). Also a note in the new figure showing the tensile test specimens was deleted. The note referred to grinding the reinforcement from the weld tensile test specimen.

Under nondestructive testing the term "electromagnetic" replaced "eddy-current". Also, the rejection limit for DSAW imperfections was changed to require investigation of any imperfection that produced an indication more than ⅓ as great as that produced by the reference standard, and one new reference standard was included for ultrasonic and electromagnetic inspection.

## Twelfth Edition Through Twenty-Fourth Edition

The next thirteen editions contained relatively few though not always minor changes. Briefly they were as follows.

**Twelfth Edition (March, 1965).** An alternative tensile test configuration was added. The diameter tolerance could be applied to the inside diameter instead of the outside diameter if the purchaser and manufacturer agreed. A provision which previously allowed the manufacturer with the purchaser's permission to use any nondestructive inspection method for any seam type was deleted, forcing radiological methods to be used exclusively for DSAW seams and ultrasonic, electromagnetic, and magnetic particle methods to be used with ERW and flash welded seams. The ultrasonic method was also permitted for DSAW seam inspection.

**Thirteenth Edition (March, 1966).** Grade X60 pipe was accepted in this edition except for use in ERW pipe (the latter could be made with Grade X60 by agreement only). With the addition of Grade X60, chemical requirements for vanadium and columbium were added. The yield strength for Grade X60 was 60,000 psi and the tensile strength was 75,000 for pipes 18-inch and under or 78,000 for pipes 20 inches and up as long as the wall thickness was less than or equal to 0.375 inch.

Under dimensions, a straightness limit requirement of 0.2 percent deviation was added; and the diameter and wall thickness tolerances were altered slightly. The diameter tolerance for cold expanded, welded pipe was changed from ±1 percent to +0.75, -0.25. For other types it remained ±1 percent. The wall thickness tolerance lower limit was changed for welded pipe of 20 inches and up from -10 to -8 percent.

The format of the document was changed. The old Section 7 (PIPE ENDS and COATINGS) was deleted. The new Section 7 was "NONDESTRUCTIVE INSPECTION". The old Section 8 (MARKINGS) became "WORKMANSHIP, VISUAL INSPECTION, AND REPAIR" the old Section 9 (INSPECTION AND REPAIR) became "MARKING AND COATING". None of these changes involved any significant change in the substance of the standard; paragraphs were merely moved to accommodate the new arrangement. However, under "workmanship" laminations of ¼ inch or more appearing at a bevel were designated as injurious requiring cutting back until they were eliminated.

**Fourteenth Edition (March, 1967).** In this edition, scope changes were made amounting to mostly "boiler plate". Grade X65 was added, however, and the prohibition of Grade X60 for ERW pipe was deleted so ERW pipe could henceforth be made from either Grade X60 or X65 skelp. The yield and tensile strengths for Grade X65 were set at 65,000 and 80,000 psi, respectively, and titanium was added to the chemical requirements.

Under the "Process" section, a requirement was added for post-weld heat treatment of ERW seams. They were to be heated either to 1000°F or to a level which would assure the absence of untempered martensite. Other changes included deleting the alternate tensile specimen configuration, the addition of a special end condition requirement for plain end pipe to be used with couplings, a change in the diameter tolerance for pipe less than 20-inch outside diameter from ±1

percent to ±0.75 percent, and some new requirements on
workmanship. Under the latter the repair of a sharp bottom gouge
in a dent by grinding was permitted, a limit of 0.060 inch was
set for offset edges at an ERW seam, and requirements were added
for inside trim and outside bead height.

**Fifteenth Edition (March, 1968).** A second value of
tensile strength of 77,000 psi was added for Grade X65. It
applied to all pipe 18-inch outside diameter or less whereas the
80,000 psi value was to be henceforth applicable to material
20-inch outside diameter and up with a wall thickness of
0.375-inch or less. The yield/tensile ratio for Grade X65 in
thicknesses of 0.375-inch or more was set at 0.90. Grade X56 was
added. Its minimum yield strength was 56,000 psi; its two
minimum values of tensile strength were 71,000 and 75,000 psi for
the sizes described above. All elongation values were moved to
an appendix table. A provision was added requiring tensile tests
to be conducted at strain rates in accordance with ASTM 370.

The upper wall thickness tolerances for pipe sizes of
20 inches and up were changed (from +15 to +17.5 percent for
seamless pipe and from +15 to +19.5 for welded pipe). The
reference standards for ultrasonic and electromagnetic inspection
were incorporated into a table and some figures in place of text.

**Sixteenth Edition (April, 1969).** The format of the
"process" section was altered such that definitions for all types
of pipe were included (the same definitions as used previously in
various places in the standard). A paragraph on heat treatment
was added, permitting API line pipe to be as-rolled, normalized,
normalized and tempered, or quenched and tempered. New diameters
included 44-inch outside diameter and the sizes 2⅜ through
4 inches. Cold die stamping was prohibited for pipe that was not

to be subsequently heat treated, although die stamping was considered "hot" not cold if done while the material was at a temperature in excess of 200°F.

Perhaps the most significant change is the Sixteenth Edition was the addition of supplementary requirements. A new Appendix (Appendix E) was created for Supplementary Requirements. Supplementary requirements were included as follows:

SR-3 Color Identification (Different paint colors for different grades)

SR-4 Nondestructive Inspection of Seamless Line Pipe

SR-5 Charpy Impact Testing on Welded Pipe
20-inch diameter or larger, Grade X52 or higher

SR-6 Drop Weight Tear Testing on Welded Pipe
20-inch Diameter or Larger, Grade X52 or Higher.

Any or all such requirements could be incorporated by agreement between the purchaser and the manufacturer into the purchase agreement. Along with these, a paragraph was added in Section 4 alluding to fracture toughness testing based on SR-5 or SR-6.

**Seventeenth Edition (April, 1970).** Bessemer steel was deleted and the chemical requirements table was simplified. Sizes to 48-inch outside diameter were added.

**Eighteenth Edition (April, 1971).** A minor change involved the addition of a table of taper angles for machining (or back-beveling) thick seamless pipe. Hard spots of 2 inches or more in diameter and hardnesses of 35 Rockwell C or more were listed as injurious defects.

**Supplement to the Eighteenth Edition (April, 1972).** A provision was added to accept pipe with two longitudinal DSAW seams 180 degrees apart for sizes above 36-inch outside diameter.[*] A figure was included to illustrate the ring tensile test. Several minor wording changes were made primarily to better define the various kinds of defects and to make the repair methods more specific. The most noticeable change was the addition to the mandatory weld ductility test for ERW and flash-welded pipe. The test involved flattening a ring of pipe between two plates with the weld at 90 degrees. Acceptable performance was to be judged in terms of a calculated plate separation that has to be reached before cracks of more than 1/8-inch appeared at the outside diameter surface. One test per lot of 200 was required for 12¾-inch outside diameter pipe and smaller, and one test per lot of 100 was required for 14-inch outside diameter pipe and larger. It was stated that the weld ductility test could serve as one of the required flattening tests.

---

[*] "The impetus for double longitudinal seam pipe arose in the early 1970's when Alyeska sought out sources of supply for the line pipe for the Trans Alaska pipeline system. The proposed diameter, 48", was larger than had been produced in North America. No US manufactured had the capability to roll and/or ship plate wide enough to make single seam 48-inch pipe, so double seam pipe seemed to be a solution. At least five manufacturers — U.S. Steel, Bethlehem, Republic, Kaiser, and A.O. Smith — expressed an interest in this possibility, and the result was the modification of the API Specification to permit two seams. One manufacturer wanted the seams at 90 degrees, but the requirement to position the seams at the neutral axis during field bending made it necessary for them to be at 180 degrees to one another. Unresolved was the question of whether all forming had to precede welding, or if one weld could be made prior to forming. The double seam concept was approved but apparently little if any pipe was produced" [E. L. Von Rosenberg, private communication, 1996].

**Nineteenth Edition (March 1973).** Grade X70 pipe was added on a tentative basis (Monogram not permitted) and the chemistry and tensile requirements were added to the tables. Grade X70 had a minimum yield strength of 70,000 psi and a minimum tensile strength of 82,000 psi.

**Supplement to the Nineteenth Edition (March, 1974).** Requirements were added for fluoroscopic operator qualifications. Arc burns were added to the list of injurious defects but contact marks from ERW electrodes were excluded from the definition. A new drawing was created for the guided bend test jig.

**Twentieth Edition (March 1976).** An optional reduced wall thickness specimen was permitted for guided bend tests of submerged-arc welded materials of 0.750-inch wall thickness and up. A revised diameter tolerance was included for pipes with diameters above 36-inch: nonexpanded pipe ±1.00 percent, cold expanded pipe +¼ inch, -⅛ inch. Under nondestructive inspection the ASME penetrameter was replaced with either the API standard penetrameter or the ISO wire penetrameter. Repair methods were specified for arc burns.

**Supplement to the Twentieth Edition (March, 1976).** The size table was expanded to include diameters of 52, 56, 60, and 64 inches. Grade X70 was fully accepted.

**Twenty-First Edition (March, 1977).** The scope was expanded to include "through-the-flowline (TFL) pipe and a supplementary requirement, SR-7, was added to provide additional requirements for such pipe. Under the inside diameter flash trim requirements, "depth of groove" was to be measured in comparison

to the wall thickness 1-inch from the bondline.  The adjective
"injurious" was deleted where used as a modifier for "defect".

**Twenty-Second Edition (March, 1978).**  Gas metal-arc
welding (MIG) was included as an acceptable seam welding process.
Allowable chemical analyses methods were stated for the first
time.  The methods were:  emission spectroscopy, X-ray emission,
atomic absorption, combustion, and wet.  Sample preparation was
to be done in accord with ASTM E-59 and testing was to be done in
accord with ASTM A-350.  The term "product analysis" replaced the
formerly used "check analysis".  Supplementary requirement, SR8,
was added for "Fracture Toughness Testing of Line Pipe" based on
the Charpy V-notch absorbed energy level.

**Twenty-Third Edition (March, 1980).**  A combination gas
metal-arc  and submerged-arc welded seam process was included.
The statement allowing the use of high strength low alloy steel
as a material was deleted.  A caution was added under chemical
requirements to the effect that certain elements even if used in
allowed amounts could adversely affect weldability.

**Supplement to the Twenty-Third Edition (March, 1981).**
The round bar tensile specimen was allowed as a tensile test
method.  The specimen was to be the largest possible transverse
specimen removable without flattening the pipe.  A statement was
added to disallow hot flattening, artificial aging, or heat
treating for any type of tensile test specimen.  Other changes
were minor.

**Twenty-Fourth Edition (March, 1982).**  This edition
incorporated the changes to the Twenty-Third Edition made in the
Supplement.  No other obvious changes were made.  This was the

last edition of the 5LX Specification. Henceforth, all grades of line pipe including the X grades were covered by the 5L Specification. The first of the "Combined" specification appeared in March, 1983.

## API Specification 5LS (1965-1982)

In 1965 the first tentative specification for "Spiral-Weld" line pipe was issued. The Second Edition appeared in 1967, and it was adopted as a "standard". Twelve editions in all were published with the last (Twelfth Edition) appearing in March, 1982. After that edition the requirements for spiral weld pipe were embodied in the 5L specification. To illustrate what these documents covered and to prepare for the discussion of the Thirty-Third Edition of Specification 5L (which combined 5L, 5LX, and 5LS), two editions of the 5LS specification, the Eighth and Twelfth, are discussed below.

### Eighth Edition

The Eighth Edition of the 5LS specification was dated March, 1975. It looked very much like the 5LX specification of the same vintage (the Nineteenth Edition of 5LX). It had the same number of sections and appendices covering the same topics. It permitted spiral-weld pipe to be made by either the ERW process or the DSAW process. The principal differences were as follows.

- The 5LS specification covered Grades A and B as well as the X grades.

- Skelp end welds were addressed in 5LS. They could be ERW, DSAW, or Flash-welded.

- A provision was included in 5LS limiting skelp width to not less than 0.8 nor more than 3.0 times the OD of the pipe.

- In the Eighth Edition of 5LS cold expansion was not permitted.

- When supplemented in March of 1976 the 5LS specification covered pipe diameters up to 80 inches.

## Twelfth Edition

The Twelfth Edition of the 5LS specification was dated March 1982. The only noticeable differences between it and the Eighth Edition were that in the meantime Grade X70 was fully accepted and cold expansion was now an acceptable procedure. The Twelfth Edition of API Specification 5LS was the last. When the Thirty-Third Edition of API Specification 5L was issued in March 1983, it covered all grades of all kinds of line pipe including spiral weld pipe.

## API Specification 5LU

API Specification 5LU was first issued as a tentative specification in 1973. It covered the requirements for heat-treated line pipe in sizes 6⅝-inch and larger in grades U80 through U100. It never became a standard and was discontinued by 1987.

It is doubtful that much U-grade pipe was ever produced or used. Research studies conducted at Battelle indicated that heat-treated carbon steels would likely be susceptible to environmental cracking problems if used in a cathodically-protected soil environment.

## API 5L Specifications (1983-1995)

### Thirty-Third Edition

The Thirty-Third Edition of API Specification 5L was issued in March, 1983. It incorporated three previously separate specifications: 5L for Grades A25, A and B only, 5LX for the high-test (X grades) of line pipe, and 5LS for spiral-weld pipe of all grades. The new document was a composite of these three and it eliminated an enormous amount of duplication formerly present with the three separate documents. The three scopes just mentioned were merged. The format included the same eleven sections found in the previous 5L specifications, two more than found in 5LX and 5LS because of the threading and coupling of standard threaded line pipe. The number and subject matter of the eight appendices was virtually unchanged as the three former specifications had generally the same material in their appendices. There were substantial changes, however, and these are discussed below.

**Process of Manufacturing and Material (Section 2).** This section incorporated all of the processes formerly recognized by the three separate documents except two. As of this edition, neither spiral-weld made by the electric resistance welding process nor flash-welding were recognized as API-approved materials. As before, the steelmaking processes allowed were basic oxygen, open-hearth, and electric-furnace. The postweld heat-treatment specified under electric-welding was extended to Grades A25 and A made by that process. To simplify the new combined document a table of processes was created which was essentially as follows.

| Process | Grade A-25 | Grades A and B | X42-X70 |
|---|---|---|---|
| Seamless | X | X | X |
| Butt-Weld[a] | X | | |
| Straight-Seam Electric Weld[b] | X | X | X |
| Straight-Seam DSAW | | X | X |
| Straight-Seam MIG | | X | X |
| Straight-seam comb MIG & DSAW | | | X |
| Spiral-seam[c] | | X | X |
| Double seam[d] | | X | X |

(a) including continuous-weld
(b) ERW and electric induction only (flash-weld no long recognized)
(c) 4½-inch OD and up, DSAW only , skelp width-to-diameter range 0.8 to 3.0
(d) 36-inch and up, DSAW, MIG, or Comb. DSAW & MIG

**Other Changes.** The substance of the remaining sections of the Thirty-Third combined edition was a composite of the three formerly separate specifications. Two new figures were added, however, which significantly simplified the process of understanding the tensile test requirements and the flattening test requirements. The addition of two new tables, one covering numbers of tests and one covering repair procedures also contributed to a more comprehensible document. One test requirement appeared to have been deleted, and that was the skelp end weld test for spiral-weld pipe. Also, on the figure for tensile test specimens a note was added prohibiting hot flattening, artificial aging or heat treatment of specimens.

## Thirty Fourth through Forty-First Editions (1984-1995)

The next eight editions of API Specification 5L contained few changes of substance, but as they evolved they clearly showed the effects of an increasing emphasis on quality assurance. This series of specifications brings us to the present time. Briefly the key features of each are as follows.

**Thirty-Fourth Edition (May 31, 1984)**

- ERW pipe post-weld heat treatment

    -minimum of 1000°F required for grades higher than Grade X42

    -for Grades A25, A, B and X42 either a heat treatment to 1000°F or assurance of no untempered martensite

- Flattening test required at crop ends of any weld stop in a continuous ERW production line. These tests could be substituted for the two intermediate tests

- The requirement for striking the pipe near the weld at each end with a 2-lb hammer during the hydrostatic test hold period was deleted.

**Thirty-Fifth Edition (May 31, 1985)**

- Grade X80 added

    -minimum yield strength 80,000 psi

    -minimum tensile strength 90,000 psi

    -maximum tensile strength 120,000 psi

    -yield/tensile ratio $\leq$ 0.93 if cold expanded

    -fracture toughness tests mandatory (SR8 and SR5 or SR8 and SR6)

- The paint mark SPEC 5L replaced the API Monogram

- Supplement 1, May 31, 1986 introduced SR14: a supplementary requirement to calculate end load compensation for hydrostatic test pressures in excess of 90 percent of SMYS.

**Thirty-Sixth Edition (June 30, 1987)**

* Format of cover changed

* Practice of denoting changes by means of black bar in margin initiated

* Last edition to provide list of manufacturers.

**Thirty-Seventh Edition (May 31, 1988)**

* Table of contents expanded to include subsection titles

* Forward incorporated into scope section

* Policy moved out of scope section

* Reference standards table added

* Where two tensile strength values had existed for certain X grades, one for pipe under 20-inch OD and one for pipe 20-inch OD and up, the higher value (for 20-inch OD and up) was deleted and the lower minimum became the required value irrespective of diameter

* SR5 was divided into two segments

  -SR5A, requirements for shear area

  -the all heat average was raised from 50 percent shear area to 80 percent shear area

  -SR5B, requirements for absorbed energy

  -SR8 was deleted.

**Thirty-Eighth Edition (May 1, 1990)**

* Provisions introduced to specify retention of records and certification

* SR15 introduced: a supplementary requirement for test certificates covering

  -size, grade, process of manufacture and heat treatment

-chemical analyses

-tensile data

-fracture toughness test results

-hydrostatic test pressure and duration

-nondestructive inspection methods

-post weld heat treatment for ERW pipe

-results of any supplemental testing

- The process definitions (i.e., seamless, ERW, DSAW, spiral-weld, etc.) were changed

  -spiral-weld pipe became helical seam SAW pipe

  -ERW was referred to as electric-welded or electric-induction welded pipe

  -butt-welded pipe became continuous-welded pipe

- The yield/tensile ratio for cold expanded pipe of any grade was raised to 0.93

- The term "manipulation tests" was introduced to cover guided bend and tensile elongation tests of SAW and GMAW seams

- The tolerances on lengths were changed


**Thirty-Ninth Edition (June 1, 1991)**

- ERW pipe post-weld heat treatment

  -normalizing required for grades higher than Grade X42

  -for Grades A25, A, B and X42 either normalizing or assurance of no untempered martensite

- Appendix A changed - Welding of jointers now covered by API Standard 1104

- Appendix B changed - Repair procedure qualification and welder qualification defined in a format like API Standard 1104

## Fortieth Edition (November 1, 1992)

- Chemical requirements

  -Phosphorus limit lowered from 0.04 to 0.03

  -Sulfur limit lowered from 0.05 to 0.03

- SR17 introduced: a supplementary requirement for nondestructive inspection of the seams of ERW pipe

- SR18 introduced: a carbon equivalent limit of 0.43 based on $C + \dfrac{Mn}{6} + \dfrac{Ni + Cu}{15} + \dfrac{Cr + Mo + V}{15}$

- Pipe body laminations of a certain size were declared to be defects

## Forty-First Edition (April 1, 1995)

- Radical reorganization with no significant change in substance

  -Expanded from 11 to 12 sections

  -Expanded from 8 to 10 appendices

  -Size changed from 8 x 10½ to 8½ by 11

  -Expanded from 94 to 119 pages but larger print size used

- Tapered ends allowed for Charpy V-notch specimens in SR5.

## Table 9-1. Summaries of API Specifications 5L

| Edition | Date | Materials | Processes | Grades | C Max | Mn Max | Mn Min | P Max | P Min | S Max | Cb Min | V Min | Ti Min | Tensile Str. psi | Yield Str. psi | Elong. | Hydrostatic Test P = 2 S t/D, S ≤2500 psig | Comments |
|---|---|---|---|---|---|---|---|---|---|---|---|---|---|---|---|---|---|---|
| 1st A P I Stds No 5-L | Jan 1928 | Bessemer Steel, Open Hearth Steel, Electric Steel, Wrought Iron, Open Hearth Iron | Butt-welded ≤3" OD, Lap-welded, Seamless (O H or electric steel only) | Bessemer Welded | -- | 0.60 | 0.30 | 0.11 | -- | 0.065 | -- | -- | -- | 50,000 | 30,000 | 20 | 14,000-16,000 | |
| | | | | O H Welded Class I | -- | 0.60 | 0.30 | 0.045 | -- | 0.06 | -- | -- | -- | 45,000 | 25,000 | 22 | 14,000-16,000 | |
| | | | | O H Welded Class II | -- | 0.60 | 0.30 | 0.08 | 0.045 | 0.06 | -- | -- | -- | 48,000 | 28,000 | 20 | 14,000-16,000 | |
| | | | | Seamless Grades A | -- | 0.60 | 0.30 | 0.04 | -- | 0.06 | -- | -- | -- | 48,000 | 30,000 | 40 | 14,000-16,000 | |
| | | | | B | -- | 1.50 | 0.35 | 0.04 | -- | 0.06 | -- | -- | -- | 70,000 | 40,000 | 25 | 18,000-20,000 | |
| | | | | C | -- | 1.50 | 0.35 | 0.04 | -- | 0.06 | -- | -- | -- | 75,000 | 45,000 | 20 | 18,000-20,000 | |
| | | | | Wrought Iron | | | | | | | | | | 42,000 | 24,000 | 12 | 12,500-14,000 | |
| | | | | Open Hearth Iron | | | | | | | | | | 42,000 | 24,000 | 20 | 12,500-14,000 | |
| 2nd A P I Stds No 5-L | Jan 1929 | Same | Same | Same | | | | Same | | | | | | Same except for 18% elong for Bessemer welded | | | Same | |
| 3rd A P I Stds No 1 Supplement No 1 Feb 1931 | Jan 1930 | Same | Same | Bessemer Welded | -- | 0.60 | 0.30 | 0.11 | -- | 0.065 | -- | -- | -- | 50,000 | 30,000 | 18 | 14,000-16,000 | Adds Appendix B A P I Line Pipe Licensees |
| | | | | O H Welded Class I | -- | 0.60 | 0.30 | 0.045 | -- | 0.06 | -- | -- | -- | 45,000 | 25,000 | 22 | | |
| | | | | O H Welded Class II | -- | 0.60 | 0.30 | 0.08 | 0.045 | 0.06 | -- | -- | -- | 48,000 | 28,000 | 20 | | |
| | | | | Seamless Grades A | -- | 0.60 | 0.30 | 0.04 | -- | 0.06 | -- | -- | -- | 48,000 | 30,000 | 40 | | |
| | | | | (for field welding)B | 0.30 | 1.50 | 0.35 | 0.04 | -- | 0.06 | -- | -- | -- | 65,000 | 38,000 | 25 | | |
| | | | | C | * | 1.50 | 0.35 | 0.04 | -- | 0.06 | -- | -- | -- | 75,000 | 45,000 | 20 | | |
| | | | | Wrought Iron | | | | | | | | | | 42,000 | 24,000 | 20 | | |
| | | | | Open Hearth Iron | | | | | | | | | | 42,000 | 24,000 | 12 | | |
| 4th | Jul 1931 | Open hearth iron permitted for seamless | Electric welded pipe added (open hearth steel); adds "furnace" welded to butt-welded lap-welded pipe distinguish from electric welded | First mention of transverse tensile for electric welded pipe | | | | Changed in supplement to be as shown in Fifth Edition | | | | | | Changed in supplement to be as shown in Fifth Edition | | | Changed in supplement to be as shown in Fifth Edition | |
| 5th A P I Stds No 5-L | Jan 1934 | Same | Same | Bessemer Welded | -- | 0.60 | 0.30 | 0.11 | -- | 0.065 | -- | -- | -- | 50,000 | 30,000 | 18 | 14,000-16,000 | Plain-end pipe bevels changed from 45 to 37¼ |
| | | | | O H Welded Class I | -- | 0.60 | 0.30 | 0.045 | -- | 0.06 | -- | -- | -- | 45,000 | 25,000 | 22 | 14,000-16,000 | |
| | | | | O H Welded Class II | -- | 0.60 | 0.30 | 0.08 | 0.045 | 0.06 | -- | -- | -- | 48,000 | 28,000 | 20 | 14,000-16,000 | |
| | | | | Seamless Grades A | -- | 0.60 | 0.30 | 0.04 | -- | 0.06 | -- | -- | -- | 48,000 | 30,000 | 30 | 14,000-16,000 | Marking change |
| | | | | B | 0.30 | 1.50 | 0.35 | 0.045 | -- | 0.06 | -- | -- | -- | 60,000 | 35,000 | 25 | 16,000-18,000 | L-Furnace welded |
| | | | | C | * | 1.50 | 0.35 | 0.04 | -- | 0.06 | -- | -- | -- | 75,000 | 45,000 | 20 | 18,000-20,000 | E-Electric welded |
| | | | | Electric Welded Grade B | 0.30 | 0.70 | 0.30 | 0.045 | -- | 0.06 | -- | -- | -- | 60,000 | 35,000 | 25 | 16,000-18,000 | S-Seamless |
| | | | | Wrought Iron | | | | | | | | | | 42,000 | 24,000 | 20 | 14,000-12,500 | I-Wrought iron OH iron |
| | | | | Open Hearth Iron | | | | | | | | | | 42,000 | 24,000 | 12 | | |
| 6th A P I Stds No 5-L | Aug 1935 | Bessemer Steel, Open-Hearth Steel, Electric Steel, Wrought Iron, Open-Hearth iron | Butt-welded ≤3", Lap-welded ≥ 3", Seamless (O H or electric steel only), Same | Bessemer Welded, O H Welded Class I, O H Welded Class II, Seamless Grade A, B, C, Electric Welded Grade B, Wrought Iron, Open Hearth Iron Seamless or Welded | | | | Same | | | | | | Same | | | Same | Definition of wrought iron added Supplement No 3 (Spetember, 1938) adds provision for metric units and enhanced purchaser inspection rights. Supplement No 6 (September, 1939) changed definition of yield strength to 0.5% total strain |
| 7th | Apr 1940 | Same | Same | Same | -- | 0.60 | 0.30 | 0.11 | -- | 0.065 | -- | -- | -- | 50,000 | 30,000 | 18 | Same | |

## Table 9-1. Summaries of API Specifications 5L

| Edition | Date | Materials | Processes | Grades | C Max | Mn Max | Mn Min | P Max | P Min | S Max | Cb Min | V Min | Ti Min | Tensile Str. psi | Yield Str. psi | Elong | Hydrostatic Test — S | Comments |
|---|---|---|---|---|---|---|---|---|---|---|---|---|---|---|---|---|---|---|
| | API Stds No 5-L | | | | | | | | | | | | | 45,000 | 25,000 | 22 | | Separate chemistry for electric welded Grade B dropped. A S A pipe thickness schedule included for information only. Supplement No 2-Plain-end bevel changed from 37¼ to 30° Phosphorus content for Grades A and C changed from 0 04 to 0 045. |
| | | | | | -- | 0.60 | 0.30 | 0.045 | -- | 0.06 | -- | -- | -- | 48,000 | 28,000 | 20 | | |
| | | | | A | -- | 0.60 | 0.30 | 0.045 | 0.045 | 0.06 | -- | -- | -- | 48,000 | 30,000 | 30 | 18,000 | |
| | | | | B | 0.30 | 1.50 | 0.35 | 0.045 | -- | 0.06 | -- | -- | -- | 60,000 | 35,000 | 25 | 21,000 | |
| | | | | C | • | 1.50 | 0.35 | 0.045 | -- | 0.06 | -- | -- | -- | 75,000 | 45,000 | 20 | 27,000 | |
| | | | | | 0.30 | | | | | | | | | 42,000 | 24,000 | 12 | 14,400 | |
| | | | | | • | | | | | | | | | 42,000 | 24,000 | 20 | 14,400 | |
| 8th | May 1942 API Stds No 5-L | Bessemer Steel | Butt-welded ≤3" | Bessemer Welded | -- | 0.60 | 0.30 | 0.11 | -- | 0.065 | -- | -- | -- | 50,000 | 30,000 | 18 | | Test pressures revised, sizes and weights reviesed revised elongations for Grades A and B--put in table by w t Paint stenciling added to die-stamping. |
| | | Open-Hearth Steel | Lap-welded ≥ 3" | O H Welded Class I | -- | 0.60 | 0.30 | 0.045 | -- | 0.06 | -- | -- | -- | 45,000 | 25,000 | 22 | | |
| | | Electric Steel / Wrought Iron | Seamless / Electric Weld | O H Welded Class II** ; Seamless or Electric Welded Grade | -- (new) | 0.60 | 0.30 | 0.08 | 0.045 | 0.06 | -- | -- | -- | 48,000 | 28,000 | 20 | | |
| | | Open-Hearth Iron | | Grade A | -- | 0.90 | 0.30 | 0.045 | -- | 0.06 | -- | -- | -- | 48,000 | 30,000 | 30 | 18,000 | |
| | | | | B | 0.30 | 1.50 | 0.35 | 0.045 | -- | 0.06 | -- | -- | -- | 60,000 | 35,000 | 25 | 21,000 | |
| | | | | C | • | 1.50 | 0.35 | 0.045 | -- | 0.06 | -- | -- | -- | 75,000 | 45,000 | 20 | 27,000 | |
| | | | | Wrought Iron | -- | | | | | | | | | 42,000 | 24,000 | 12 | 14,400 | |
| | | | | Open-Hearth Iron | -- | | | | | | | | | 42,000 | 24,000 | 20 | 14,400 | |
| 9th | Aug 1944 API Stds No 5-L | Bessemer Steel | Butt-welded ≤3" ; Lap-welded ≥ 3" | Furnace Butt and Lap Welded ; Bessemer | -- | 0.60 | 0.30 | 0.11 | -- | 0.065 | -- | -- | -- | 50,000 | 30,000 | 18 | 18,000 | |
| | | Open-Hearth Steel | Seamless (O H or electric steel only) | Electric furnace — Grade B | -- | 0.60 | 0.30 | 0.045 | -- | 0.06 | -- | -- | -- | 45,000 | 25,000 | 22 | 21,000 | |
| | | Electric Steel / Wrought Iron | Electric Weld | C | -- | 0.60 | 0.30 | 0.08 | 0.045 | 0.06 | -- | -- | -- | 25,000 | 45,000 | 22 | 27,000 | |
| | | Open-Hearth iron | | O H Welded Class I ; O H Welded Class II** ; Seamless — Grade A | -- | 0.60 | 0.30 | 0.045 | -- | 0.06 | -- | -- | -- | 28,000 | 48,000 | 20 | | |
| | | | | Bessemer — Grade B | 0.30 | 0.90 | 0.35 | 0.045 | -- | 0.06 | -- | -- | -- | 60,000 | 35,000 | (a) | 18,000 | |
| | | | | C | • | 1.50 | 0.35 | 0.045 | -- | 0.06 | -- | -- | -- | 75,000 | 45,000 | 20 | 27,000 | |
| | | | | Electric Furnace — Grade A | -- | 0.90 | 0.30 | 0.045 | -- | 0.06 | -- | -- | -- | 48,000 | 30,000 | (a) | 18,000 | |
| | | | | B | 0.3 | 1.50 | 0.35 | 0.045 | -- | 0.06 | -- | -- | -- | 60,000 | 35,000 | (a) | 21,000 | |
| | | | | C | • | 1.50 | 0.35 | 0.045 | -- | 0.06 | -- | -- | -- | 75,000 | 45,000 | 20 | 27,000 | |
| | | | | Open Hearth — Grade A | -- | 0.90 | 0.30 | 0.045 | -- | 0.06 | -- | -- | -- | 48,000 | 30,000 | (a) | 18,000 | |
| | | | | B | 0.3 | 1.50 | 0.35 | 0.045 | -- | 0.06 | -- | -- | -- | 60,000 | 35,000 | (a) | 21,000 | |
| | | | | C | • | 1.50 | 0.35 | 0.045 | -- | 0.06 | -- | -- | -- | 75,000 | 45,000 | 20 | 27,000 | |
| | | | | Electric Welded ; Bessemer — Grade B | 0.30 | 0.90 | 0.35 | 0.11 | -- | 0.06 | -- | -- | -- | 60,000 | 35,000 | (a) | 21,000 | |
| | | | | C | • | 1.50 | 0.35 | 0.11 | -- | 0.06 | -- | -- | -- | 75,000 | 45,000 | 20 | 27,000 | |
| | | | | Electric Furnace — Grade A | -- | 0.90 | 0.30 | 0.045 | -- | 0.06 | -- | -- | -- | 48,000 | 30,000 | (a) | 18,000 | |
| | | | | B | 0.3 | 1.50 | 0.35 | 0.045 | -- | 0.06 | -- | -- | -- | 60,000 | 35,000 | (a) | 21,000 | |
| | | | | C | • | 1.50 | 0.35 | 0.045 | -- | 0.06 | -- | -- | -- | 75,000 | 45,000 | 20 | | |
| | | | | Open Hearth — Grade A | -- | 0.90 | 0.30 | 0.045 | -- | 0.06 | -- | -- | -- | 48,000 | 30,000 | (a) | | |
| | | | | B | 0.30 | 1.50 | 0.35 | 0.045 | -- | 0.06 | -- | -- | -- | 60,000 | 35,000 | (a) | | |

Hydrostatic Test: $P = 2\,S\,t/D$ ≤2500 psig

## Table 9-1. Summaries of API Specifications 5L

| Edition Date | Materials | Processes | Grades | C Max | Mn Max | Mn Min | P Max | P Min | S Max | Cb Min | V Min | Ti Min | Tensile Str. psi | Yield Str. psi | Elong. | Hydrostatic Test $P = 2 S t/D$, $S$ ≤2500 psig | Comments |
|---|---|---|---|---|---|---|---|---|---|---|---|---|---|---|---|---|---|
| 10th Aug 1945 API STD 5-L | Same | Butt-welded ≤4" Lap-welded ≥ 4"; Seamless (open hearth or electric steel only); Electric Weld | Wrought Iron; Open Hearth Iom; Furnace Butt and Lap Welded; Bessemer; Electric Furance; O.H. Welded Class I; O H. Welded Class II**; Open Hearth Iron; Wrought Iron; Seamless: Bessemer Grade B, C; Electric Furance Grade A, B, C; Open Hearth Grade A, B, C; Open Hearth Iron Grade A, B, C; Electric Welded Bessemer Grade B, C; Electric Furance Grade A, B, C; Open Hearth Grade A, B, C; Open Hearth Iron | C | 1.50 | 0.35 | 0.045 | -- | 0.06 | -- | -- | -- | 75,000 42,000 42,000 | 45,000 24,000 24,000 | 20 12 20 | 27,000 14,400 14,400 | |
| Supplement No 1 Jul 1946 Supplement No 2 Sep 1947 | | | | Same | | | Same | | | | | | Same | | | Same | |
| 11th May 1949 API Std 5L | Bessemer Steel; Open Hearth Steel; Electric Steel; Wrought Iron; Open Hearth Iron | Steel Seamless Electric Welded; Seamless Electric Welded; Electric-furnace or O H; Bessemer; Lap- or Butt-welded; Electric-furnace; Open-hearth; Bessemer; Iron Lap- or Butt-Welded; Electric Furnace; Seamless or electric-welded; Open-hearth Iron; Lap- or butt-welded; Open-Hearth; Wrought | Open Hearth Iron; Seamless or Electric Welded Grade A; Electric-furnace or O H Grade A; B; C; Bessemer Grade B; C; Lap- or Butt-Welded Electric Furnace; O H Class I; O H Class II; Bessemer; Open Hearth Iron | -- 0.30 0.30 0.30 -- -- -- -- -- | 0.90 1.50 1.50 1.50 1.50 0.60 0.60 0.60 0.60 | 0.30 0.35 0.35 0.35 0.35 0.30 0.30 0.30 0.30 | 0.045 0.045 0.045 0.11 0.11 0.045 0.045 0.08 0.11 | -- | 0.06 0.06 0.06 0.06 0.06 0.06 0.06 0.06 0.065 | -- | -- | -- | 48,000 60,000 75,000 60,000 75,000 45,000 45,000 48,000 50,000 42,000 42,000 | 30,000 35,000 45,000 35,000 45,000 25,000 25,000 28,000 30,000 24,000 24,000 | -- -- -- -- -- 22 22 20 18 20 12 | 18,000 21,000 27,000 21,000 27,000 18,000 18,000 18,000 18,000 14,400 14,400 | Change in format. Double extra strong |
| 12th Mar 1951 API STD 5L Supplement No 1 Jan 1952 | Same | Same | Same | -- 0.30 * | 0.90 1.50 1.50 | 0.30 0.35 0.35 | 0.045 0.045 0.045 | -- | 0.06 0.06 0.06 | -- | -- | -- | Same | | | Same | |

# Table 9-1. Summaries of API Specifications 5L

| Edition / Date | Materials | Processes | Grades | C Max | Mn Max | Mn Min | P Max | P Min | S Max | Cb Min | V Min | Ti Min | Tensile Str. psi | Yield Str. psi | Elong. | Hydrostatic Test P = 2 S t/D, S ≤2500 psig | Comments |
|---|---|---|---|---|---|---|---|---|---|---|---|---|---|---|---|---|---|
| changes in dimensions | | | | 0.30 * | 1.50 1.50 | 0.35 0.35 | 0.11 0.11 | | 0.06 0.06 | -- -- | -- -- | -- -- | | | | | |
| 13th Mar 1954 API Std 5-L | Same | Same | Same | -- | 0.60 | 0.30 | 0.045 | | 0.06 | -- | -- | -- | | | | | |
| | | | | -- | 0.60 | 0.30 | 0.045 | | 0.06 | -- | -- | -- | | | | | |
| | | | | -- | 0.60 | 0.30 | 0.08 | 0.045 | 0.06 | -- | -- | -- | | | | | |
| | | | | -- | 0.60 | 0.30 | 0.11 | | 0.065 | -- | -- | -- | | Same | | | |
| 14th Mar 1955 API Std 5-L | Same | Same / Electric Welded | Same | | | | Same | | | | | | | Same | | Same | Grade C deleted First time check analysis tolerances listed |
| | Same | Same | Seamless | | | | | | | | | | | | | | |
| | | | Electric-Furnace or O H Grade A | 0.22 | 0.95 | -- | 0.04 | -- | 0.05 | -- | -- | -- | 48,000 | 30,000 | -- | | |
| | | | B | 0.27 | 1.35 | -- | 0.04 | -- | 0.05 | -- | -- | -- | 60,000 | 35,000 | -- | | |
| | | | Killed, deoxidized, acid-bessemer Grade B | 0.22 | 1.35 | -- | 0.1 | -- | 0.05 | -- | -- | -- | 60,000 | 35,000 | -- | | |
| | | | Electric Welded | | | | | | | | | | | | | | |
| | | | Electric furnace or O H Grade A | 0.21 | 0.95 | -- | 0.04 | -- | 0.05 | -- | -- | -- | | | | | |
| | | | B | 0.26 | 1.35 | -- | 0.04 | -- | 0.05 | -- | -- | -- | | | | | |
| | | | Killed, deoxidized, acid-bessemer Grade B | 0.21 | 1.35 | -- | 0.1 | -- | 0.05 | -- | -- | -- | | | | | |
| | | | Lap- or Butt Welded | | | | | | | | | | | | | | |
| | | | Electric-Furnace | -- | 0.6 | 0.3 | 0.045 | -- | 0.06 | -- | -- | -- | 45,000 | 25,000 | 22 | | |
| | | | O H Class I | -- | 0.6 | 0.3 | 0.045 | 0.045 | 0.06 | -- | -- | -- | 45,000 | 25,000 | 22 | | |
| | | | O H Class II | -- | 0.6 | 0.3 | 0.08 | -- | 0.06 | -- | -- | -- | 48,000 | 28,000 | 20 | | |
| | | | Bessemer | -- | 0.6 | 0.3 | 0.11 | -- | 0.065 | -- | -- | -- | 50,000 | 30,000 | 18 | | |
| | | | Seamless, E W, L W. | | | | | | | | | | 42,000 | 24,000 | 20 | | |
| | | | B W O H Iron | | | | | | | | | | 42,000 | 24,000 | 12 | | |
| | | | L W. & B W Wrought Iron | | | | | | | | | | | | | | |
| 15th Mar 1956 API. Std 5L. | Same | Same | SEAMLESS | | | | | | | | | | | Same | | Same | |
| | | | Electric-furnace, O H or killed, deoxidized basic-Bessemer*** Grade A | 0.20 | 0.95 | -- | 0.04 | -- | 0.05 | -- | -- | -- | | | | | |
| | | | B | 0.27 | 1.35 | -- | 0.04 | -- | 0.05 | -- | -- | -- | | | | | |
| | | | Killed, deoxidized, acid-Bessemer, or killed, deoxidized, basic-Bessemer*** Grade B | 0.22 | 1.35 | -- | 0.1 | -- | 0.05 | -- | -- | -- | | | | | |
| | | | ELECTRIC WELDED Electric-furnace, O H or killed, deoxidized basic-Bessemer*** Grade A | 0.21 | 0.95 | -- | 0.04 | -- | 0.05 | -- | -- | -- | | | | | |
| | | | B | 0.26 | 1.35 | -- | 0.04 | -- | 0.05 | -- | -- | -- | | | | | |
| | | | Killed, deoxidized, acid-Bessemer, or killed, deoxidized, basic-Bessemer*** Grade B | 0.21 | 1.35 | -- | 0.10 | -- | 0.05 | -- | -- | -- | | | | | |
| | | | LAP- OR BUTT WELDED Electric-Furnace | -- | 0.60 | 0.30 | 0.045 | 0.045 | 0.06 | -- | -- | -- | | | | | |
| | | | O H Class I | -- | 0.60 | 0.30 | 0.045 | 0.045 | 0.06 | -- | -- | -- | | | | | |
| | | | O H Class II** | -- | 0.60 | 0.30 | 0.08 | 0.045 | 0.06 | -- | -- | -- | | | | | |
| | | | Bessemer | -- | 0.6 | 0.3 | 0.11 | -- | 0.065 | -- | -- | -- | | | | | |
| | | | IRON Seamless or E W | | | | | | | | | | | | | | |

## Table 9-1. Summaries of API Specifications 5L

| Edition | Date | Materials | Processes | Grades | C Max | Mn Max | Mn Min | P Max | P Min | S Max | Cb Min | V Min | Ti Min | Tensile Str. psi | Yield Str. psi | Elong. | Hydrostatic Test P = 2 S t/D, S ≤2500 psig | Comments |
|---|---|---|---|---|---|---|---|---|---|---|---|---|---|---|---|---|---|---|
| 16th | Apr 1957 API Std 5L. | Same | Same | Open Hearth Lap- or Butt Welded Open Hearth Wrought | | | | | | | | | | | Same | | Same | |
| | | | | SEAMLESS Electric-Furnace or O H Grade A | | | | | | | | | | | | | Same | |
| | | | | Electric-furnace, O H , or killed deoxidized basic-Bessemer | 0 22 | 0.95 | -- | 0.04 | -- | 0.05 | -- | -- | -- | 40,000 | 30,000 | -- | | |
| | | | | Grade B Killed, deoxidized, acid-Bessemer, or killed deoxidized basic-Bessemer | 0.27 | 1 35 | -- | 0.04 | -- | 0.05 | -- | -- | -- | 60,000 | 35,000 | -- | | |
| | | | | Grade B ELECTRIC WELDED Electric Furnace, O.H. | 0 22 | 1 35 | -- | 0 10 | -- | 0.05 | -- | -- | -- | | | | | |
| | | | | Grade A Electric-furnace, O H , or killed deoxidized basic-Bessemer | 0 21 | 0 95 | -- | 0.04 | -- | 0.05 | -- | -- | -- | | | | | |
| | | | | Grade B Killed, deoxidized, acid-Bessemer, or killed doxidized basic-Bessemer | 0.26 | 1 35 | -- | 0.04 | -- | 0.05 | -- | -- | -- | | | | | |
| | | | | Grade B LAP- OR BUTT WELDED Electric-Furnace | 0 21 | 1 35 | -- | 0 10 | -- | 0.05 | -- | -- | -- | | | | | |
| | | | | O H Class I | | 0.60 | 0.30 | 0.045 | -- | 0.06 | -- | -- | -- | 45,000 | 25,000 | 22 | | |
| | | | | O H Class II** | | 0.60 | 0.30 | 0.045 | -- | 0.06 | -- | -- | -- | 45,000 | 25,000 | 22 | | |
| | | | | Bessemer | | 0.60 | 0.30 | 0.08 | 0.045 | 0.06 | -- | -- | -- | 48,000 | 28,000 | 20 | | |
| | | | | IRON Seamless or E W | | 0.60 | 0.30 | 0 11 | -- | 0.065 | -- | -- | -- | 50,000 | 30,000 | 18 | | |
| | | | | Open Hearth Lap- or Butt Welded | | | | | | | | | | 42,000 | 24,000 | 20 | | |
| | | | | Open Hearth Wrought | | | | | | | | | | 42,000 42,000 | 24,000 24,000 | 20 12 | | |
| 17th | Mar 1958 | Same | Steel Seamless, electric-flash, or electric resistance welded | Seamless | | | | | | Same | | | | | Same | | Same | |
| | API Std 5L | | Electric-furnace, O H , or basic-oxygen | Electric-Furnace or O H or basic-oxygen# Grade A | | | | | | | | | | | | | | |
| | Supplement No 1 Mar 1959 | | Bessemer | Electric-Furnace, O H , basic-oxygen#, or killed deoxidized basic-Bessemer Grade B | | | | | | | | | | | | | | |
| | replaces chemical table & list of companies | | Electric induction-welded (6-5/8" and smaller, 0 280" wall thickness and less) | Grade B Killed, deoxidized, acid-Bessemer, or killed doxidized basic-Bessemer | | | | | | | | | | | | | | |
| | | | Electric-furnace, O H , ba | Grade B | | | | | | | | | | | | | | |
| | | | Lap- or Butt-Welded Electric-Furnace | Electric Welded | | | | | | | | | | | | | | |
| | | | Open-Hearth or basic oxygen | Electric-Furnace, O H , or basic oxygen# | | | | | | | | | | | | | | |
| | | | Bessemer | Grade A Electric-Furnace, O H , basic-oxygen#, or killed deoxidized basic-Bessemer | | | | | | | | | | | | | | |
| | | | O H Iron L W & B O H Seamless | Grade B Killed, deoxidized, acid-Bessemer, or killed doxidized basic-Bessemer | | | | | | | | | | | | | | |
| | | | Wrought Iron | Grade B LAP- OR BUTT WELDED Electric-Furnace | | | | | | | | | | | | | | |
| | | | | O H or basic-oxygen# Class I | | | | | | | | | | | | | | |

## Table 9-1. Summaries of API Specifications 5L

| Edition | Date | Materials | Processes | Grades | C Max | Mn Max | Mn Min | P Max | P Min | S Max | Cb Min | V Min | Ti Min | Tensile Str. psi | Yield Str. psi | Elong. | Hydrostatic Test P = 2 S t/D, S ≤2500 psig | Comments |
|---|---|---|---|---|---|---|---|---|---|---|---|---|---|---|---|---|---|---|
| 18th Supplement No 1 replaces chemical and tensile prop. | Feb 1960 Jan 1961 | Open Hearth Electric Furnace Bessemer Basic-oxygen | Seamless Wrought Steel tublar Electric Welded Electric-flash, Electric resistance or electric induction welding (6¾" and smaller, 0 280" wall thickness and less) | Seamless | | | | | | | | | | | | | | |
| | | | | Electric-Furnace or O H or basic-oxygen    Grade A | 0 22 | 0 9 | -- | 0 04 | -- | 0 05 | -- | -- | -- | Same | Same | | | Same |
| | | | | Electric-Furnace, O H, basic-oxygen, or killed deoxidized basic-Bessemer    Grade B | 0 27 | 1 15 | -- | 0 04 | -- | 0 05 | -- | -- | -- | | | | | |
| | | Acid-Oxygen Steel Wrought Iron | Lap Welded | Killed, deoxidized, acid-Bessemer, or killed deoxidized basic-Bessemer    Grade B | 0 22 | 1 15 | -- | 0 1 | -- | 0 05 | -- | -- | -- | | | | | |
| | | | | Electric Welded | | | | | | | | | | | | | | |
| | | | | Electric-Furnace, O.H., or basic oxygen    Grade A | 0 21 | 0 9 | -- | 0 04 | -- | 0 05 | -- | -- | -- | | | | | |
| | | | | Electric-Furnace, O H , basic-oxygen, or killed deoxidized basic-Bessemer    Grade B | 0 26 | 1 15 | -- | 0 04 | -- | 0 05 | -- | -- | -- | | | | | |
| | | | | Killed, deoxidized, acid-Bessemer, or killed deoxidized basic-Bessemer    Grade B | 0 21 | 1 15 | -- | 0 1 | -- | 0 05 | -- | -- | -- | | | | | |
| | | | | LAP- OR BUTT WELDED | | | | | | | | | | | | | | |
| | | | | Electric-Furnace    Class 1 | -- | 0 6 | 0 3 | 0 045 | -- | 0 06 | -- | -- | -- | | | | | |
| | | | | O H or basic-oxygen , acid-oxygen-steam, or acid-oxygen+    Class 1 | -- | 0 6 | 0 3 | 0 045 | -- | 0 06 | -- | -- | -- | | | | | |
| | | | | O H or basic-oxygen    Class II** | -- | 0 6 | 0 3 | 0 08 | 0 045 | 0 06 | -- | -- | -- | | | | | |
| | | | | Bessemer | -- | 0 6 | 0 3 | 0 11 | -- | 0 065 | -- | -- | -- | | | | | |
| | | | | Wrought | | | | | | | | | | | | | | |
| 19th API Std 5L | Mar 1962 | Same | Seamless Electric Welded Lap- and Butt-Welded Submerged arc-welded Cold Expansion (except Lap- and Butt Welded, shall be either non-expanded or cold expanded) | Seamless | | | | | | | | | | Same | | | | |
| | | | | Electric-Furnace or O H or basic-oxygen    Grade A | 0 22 | 0 9 | -- | 0 04 | -- | 0 05 | -- | -- | -- | | | | 22,500-18,000 | |
| | | | | Electric-Furnace, O.H., basic-oxygen, or killed deoxidized basic-Bessemer    Grade B | 0 27 | 1 15 | -- | 0 04 | -- | 0 05 | -- | -- | -- | | | | 26,250-21,000 | |
| | | | | Killed, deoxidized, acid-Bessemer, or killed deoxidized basic-Bessemer    Grade B | 0 222 | 1 15 | -- | 0 1 | -- | 0 05 | -- | -- | -- | | | | 26,250-21,000 | |
| | | | | Electric or Sub-merged-arc welded | | | | | | | | | | | | | | |
| | | | | Electric-Furnace, O.H., or basic oxygen    Grade A | 0 21 | 0 9 | -- | 0 04 | -- | 0 05 | -- | -- | -- | | | | 22,500-18,000 | |
| | | | | Electric-Furnace, O H , basic-oxygen, or killed deoxidized basic-Bessemer    Grade B | 0 26 | 1 15 | -- | 0 04 | -- | 0 05 | -- | -- | -- | | | | 26,250-21,000 | |
| | | | | Killed, deoxidized, acid-Bessemer, or killed deoxidized basic-Bessemer    Grade B | 0 21 | 1 15 | -- | 0 1 | -- | 0 05 | -- | -- | -- | | | | | |

## Table 9-1. Summaries of API Specifications 5L

| Edition / Date | Materials | Processes | Grades | C Max | Mn Max | Mn Min | P Max | P Min | S Max | Cb Min | V Min | Ti Min | Tensile Str. psi | Yield Str. psi | Elong. | Hydrostatic Test S ≤2500 psig ($P = 2\,S\,t/D$) | Comments |
|---|---|---|---|---|---|---|---|---|---|---|---|---|---|---|---|---|---|
| | | | **Butt Welded** Electric-Furnace | -- | 0.6 | 0.3 | 0.045 | -- | 0.05 | -- | -- | -- | | | | | |
| | | | O H or basic-oxygen, acid-oxygen-steam, or acid-oxygen **Class I** / **Class II++** | -- | 0.6 | 0.3 | 0.045 | -- | 0.06 | -- | -- | -- | | | | | |
| | | | Bessemer | -- | 0.6 | 0.3 | 0.08 | 0.045 | 0.06 | -- | -- | -- | | | | | |
| | | | Wrought Iron Lap-Welded / Wrought Iron | -- | 0.6 | 0.3 | 0.11 | -- | 0.065 | -- | -- | -- | | | | 14,400 | |
| 20th Mar 1963 API Std 5L Mar 1956 | Open-Hearth | Seamless | Seamless | Same | | | | | | | | | | | | | |
| Erratum Mar 1993 | Electric-Furnace Bessemer | Electric-Welded Submerged-Arc Weld | **Grade A** Electric-Furnace or O H or basic-oxygen | | | | | | | | | | 48,000 | 30,000 | -- | 22,500-18,000 | |
| Supplement No 1 | Basic Oxygen | Butt Welded Cold Expansion (except Butt-Welded, shall be either non-expanded or cold expanded) | **Grade B** Electric-Furnace, O H, basic-oxygen, or killed deoxidized basic-Bessemer | | | | | | | | | | 60,000 | 35,000 | -- | 26,250-21,000 | |
| Mar 1964 | Acid-oxygen-steam steel | | **Grade B** Killed, deoxidized, acid-Bessemer, or killed deoxidized basic-Bessemer | | | | | | | | | | 60,000 | 35,000 | -- | 26,250-21,000 | |
| | | | **Grade B** Electric or Sub-merged-arc welded Electric-Furnace, O H, or basic oxygen | | | | | | | | | | 48,000 | 30,000 | -- | 22,500-18,000 | |
| | | | **Grade A** Electric-Furnace, O H, basic-oxygen, or killed deoxidized basic-Bessemer | | | | | | | | | | 60,000 | 35,000 | -- | 26,250-21,000 | |
| | | | **Grade B** Killed, deoxidized, acid-Bessemer, or killed deoxidized basic-Bessemer | | | | | | | | | | 60,000 | 35,000 | -- | 26,250-21,000 | |
| | | | **Butt Welded** Electric-Furnace O H or basic-oxygen, acid-oxygen-steam, or acid-oxygen **Class I** / **Class II++** | | | | | | | | | | 45,000 | 25,000 | 22 | | |
| | | | Bessemer | | | | | | | | | | 45,000 / 48,000 / 50,000 | 25,000 / 28,000 / 30,000 | 22 / 22 / 18 | | |
| 21st Mar 1965 API Std 5L Supplement No 1 Mar 1967 | Same | Same | Same | Same | | | | | | | | | Same | | | Same | |
| 22nd Mar 1967 API Std 5L | Same | Same | Same | Same | | | | | | | | | Same | | | Same | |
| 23rd Mar 1968 API Std 5L | Open-Hearth Electric-Furnace Bessemer Basic-Oxygen | Same | Seamless Electric-Furnace or O H or basic-oxygen **Grade A** / Electric-Furnace, O H, basic-oxygen, or killed deoxidized basic-Bessemer **Grade B** | Same | | | | | | | | | Same | | | Same | |

## Table 9-1. Summaries of API Specifications 5L

| Edition | Date | Materials | Processes | Grades | C Max | Mn Max | Mn Min | P Max | P Min | S Max | Cb Min | V Min | Ti Min | Tensile Str. psi | Yield Str. psi | Elong. | Hydrostatic Test P = 2 S t/D, S ≤2500 psig |
|---|---|---|---|---|---|---|---|---|---|---|---|---|---|---|---|---|---|
| | | | | Killed, deoxidized, acid-Bessemer, or killed deoxidized basic-Bessemer **Grade B** | | | | | | | | | | | | | |
| | | | | Electric or Sub-merged-arc welded | | | | | | | | | | | | | |
| | | | | Electric-Furnace, O.H., or basic oxygen **Grade A** | | | | | | | | | | | | | |
| | | | | Electric-Furnace, O.H., basic-oxygen, or killed deoxidized basic-Bessemer **Grade B** | | | | | | | | | | | | | |
| | | | | Killed, deoxidized, acid-Bessemer, or killed deoxidized basic-Bessemer **Grade B** | | | | | | | | | | | | | |
| | | | | Butt Welded **Grade B** | | | | | | | | | | | | | |
| | | | | Electric-furnace, O.H., or basic-oxygen **Class I** | | | | | | | | | | | | | |
| | | | | **Class II**\*\* | | | | | | | | | | | | | |
| 24th Apr 1969 API Std 5L | | Same | Seamless | Bessemer | | | | | | | | | | | | | |
| | | | Electric-Weld Submerged-arc Weld (Applicable to Grades A&B only) | Electric-Furnace, O.H., or basic-oxygen | | | | | | | | | | | | | |
| | | | Butt-Weld (Applicable to Grade A25 only) | Grade A25, Class I | 0.21 | 0.60 | 0.30 | 0.045 | --- | 0.06 | --- | --- | --- | | | | |
| | | | Cold Expansion (except Butt-Welded, shall be either non-expanded or cold expanded) | Grade A25, Class II\*\* | 0.21 | 0.60 | 0.30 | 0.08 | 0.045 | 0.06 | --- | --- | --- | | | | |
| | | | | Electric-Furnace, O.H., basic-oxygen, or killed deoxidized basic-Bessemer **Grade A** | 0.22 | 0.60 | --- | 0.04 | --- | 0.06 | --- | --- | --- | | | | 22,500 18,000 |
| | | | | Killed, deoxidized, acid-Bessemer, or killed deoxidized basic-Bessemer **Grade B** | 0.27 | 1.15 | --- | 0.04 | --- | 0.05 | --- | --- | --- | | | | 26,500 21,000 |
| | | | | Electric or Sub-merged-arc welded **Grade B** | 0.22 | 1.15 | --- | 0.10 | --- | 0.05 | --- | --- | --- | | | | 26,500 21,000 |
| | | | | Grade A25, Class I(E.W. only) | 0.21 | 0.60 | 0.30 | 0.045 | --- | 0.06 | --- | --- | --- | | | | |
| | | | | Grade A25, Class II (E.W. only)\*\* | 0.21 | 0.60 | 0.30 | 0.08 | 0.045 | 0.06 | --- | --- | --- | | | | |
| | | | | **Grade A** | 0.21 | 0.90 | --- | 0.04 | --- | 0.05 | --- | --- | --- | 48,000 | 30,000 | | 22,500 18,000 |
| | | | | Electric-Furnace, O.H., basic-oxygen, or killed deoxidized basic-Bessemer **Grade B** | 0.26 | 1.15 | --- | 0.04 | --- | 0.05 | --- | --- | --- | 60,000 | 35,000 | | 26,500 21,000 |
| | | | | Killed, deoxidized, acid-Bessemer, or killed deoxidized basic-Bessemer **Grade B** | 0.21 | 1.15 | --- | 0.10 | --- | 0.05 | --- | --- | --- | | | | |
| | | | | Butt Welded | 0.21 | 0.60 | 0.30 | 0.045 | --- | 0.06 | --- | --- | --- | 45,000 | 25,000 | | 26,500 21,000 |
| | | | | Electric-furnace, O.H., or basic-oxygen **Grade A25, Class I** | 0.21 | 0.60 | 0.30 | 0.08 | 0.045 | 0.06 | --- | --- | --- | | | | |
| | | | | **Grade A25, Class II**\*\* | | | | | | | | | | Same | | | |
| 25th Apr 1970 | | Open-Hearth | Same | Seamless | | | | | | | | | | | | | |

## Table 9-1. Summaries of API Specifications 5L

| Edition Date | Materials | Processes | Grades | C Max | Mn Max | Mn Min | P Max | P Min | S Max | Cb Min | V Min | Ti Min | Tensile Str. psi | Yield Str. psi | Elong. | Hydrostatic Test P = 2 S t/D — S | ≤2500 psig | Comments |
|---|---|---|---|---|---|---|---|---|---|---|---|---|---|---|---|---|---|---|
| API Std 5L, Errata Aug 1970 | Electric-Furnace Basic-Oxygen Steel | | Grade A25, Class I | 0.21 | 0.60 | 0.30 | 0.045 | -- | 0.06 | -- | -- | -- | | | | | | |
| | | | Grade A25, Class II** | 0.21 | 0.60 | 0.30 | 0.08 | 0.045 | 0.06 | -- | -- | -- | | | | 22,500 | 18,000 | |
| | | | Grade A | 0.22 | 0.90 | -- | 0.04 | -- | 0.05 | -- | -- | -- | | | | 26,500 | 21,000 | |
| | | | Grade B | 0.27 | 1.15 | -- | 0.04 | -- | 0.05 | -- | -- | -- | | | | | | |
| | | | Electric or Submerged-arc weld | | | | | | | | | | | | | | | |
| | | | Grade A25, Class I (E.W. only) | 0.21 | 0.60 | 0.30 | 0.045 | -- | 0.06 | -- | -- | -- | | | | | | |
| | | | Grade A25, Class II(E.W. only) | 0.21 | 0.60 | 0.30 | 0.08 | 0.045 | 0.06 | -- | -- | -- | | | | 22,500 | 18,000 | |
| | | | Grade A | 0.21 | 0.90 | -- | 0.04 | -- | 0.05 | -- | -- | -- | | | | 26,500 | 21,000 | |
| | | | Grade B | 0.26 | 1.15 | -- | 0.04 | -- | 0.05 | -- | -- | -- | | | | | | |
| | | | Butt-Weld | | | | | | | | | | | | | | | |
| | | | Grade A25, Class I | 0.21 | 0.60 | 0.30 | 0.045 | -- | 0.06 | -- | -- | -- | | | | | | |
| | | | Grade A25, Class II** | 0.21 | 0.60 | 0.30 | 0.08 | 0.045 | 0.06 | -- | -- | -- | | | | | | |
| 26th Apr 1971 API STD 5L Supplement No 1 Apr. 1972 | Same | Same | Same | | | | | | Same | | | | | Same | | Same | | |
| 27th Mar 1973 API Spec 5L Supplement No 1 | Same | Seamless; Electric-Weld; Submerged-arc Weld (Applicable to Grades A & B only) | Same | | | | | | Same | | | | | Same | | Same | | |
| Mar 1974 | | Submerged-arc Weld, Double Seam (Applicable to Grades A & B in sizes larger than 36" O D ); Butt-Weld (Applicable to Grade A25 only); Cold Expansion (except Butt-Welded, shall be either non-expanded or cold expanded) | | | | | | | | | | | | | | | | |
| 28th Mar 1975 API Spec 5L Supplement No 1 Mar 1976 | Same | Same | Same | | | | | | Same | | | | | Same | | Same | | |
| 29th Mar 1977 API Spec 5L | Same | Same | Same | | | | | | Same | | | | | Same | | Same | | |
| 30th Mar 1978 API Spec 5L | Same | Seamless; Electric-Weld; Submerged-arc Weld (Applicable to Grades A & B only); Gas Metal-arc Weld (MIG) (Applicable to Grades A & B only); Double Seam Welded Pipe (Applicable to Grades A & B in sizes larger than 36" O D ); Butt-Weld (Applicable to Grade A25 only) | Seamless; Grade A25, Class I; Grade A25, Class II**; Grade A; Grade B; Electric Weld, Submerged arc Weld, or Gas Metal-arc Weld | | | | | | | | | | | | | | | |

## Table 9-1. Summaries of API Specifications 5L

| Edition Date | Materials | Processes | Grades | C Max | Mn Max | Mn Min | P Max | P Min | S Max | Cb Min | V Min | Ti Min | Tensile Str. psi | Yield Str. psi | Elong. | Hydrostatic Test P=2 S t/D, S ≤2500 psig | Comments |
|---|---|---|---|---|---|---|---|---|---|---|---|---|---|---|---|---|---|
| 31st Mar 1980 API Spec 5L Supplement No 1 Mar 1981 | | Cold Expansion (except butt-welded, shall be either non-expanded or cold expanded) | Grade A25, Class I (E W only) Grade A25, Class II(E W only)** Grade A Grade B Butt-Weld Grade A25, Class I Grade A25, Class II** | | | | | | | | | | | | | | |
| | Same | Same | Same | | | | Same | | | | | | | Same | | Same | |
| 32nd Mar 1982 API Spec 5L | Same | Same | Same | | | | Same | | | | | | | Same | | Same | |
| 33rd Mar 1983 API Spec 5L | Open-Hearth Electric-Furnace Basic-Oxygen Steel Spiral Weld skelp widths shall not be less than 0 8 or more than 0 8 or more than 3 0 times the pipe O D | Seamless Electric Weld Submerged-Arc Weld Gas Metal-Arc Weld (MIG) | Seamless Non-expanded or cold Grade A25, Class I Expanded | 0 21 | 0 60 | 0 30 | 0 045 | -- | 0 06 | -- | -- | -- | 45,000 | 25,000 | | see document, these are not the same. They are in a percentage | |
| | | Combination Gas Metal-Arc Weld | Grade A25, Class II** Grade A Grade B | 0 21 0 22 0 27 | 0 60 0 90 1 15 | 0 30 -- -- | 0 08 0 04 0 04 | 0 045 -- -- | 0 06 0 05 0 05 | -- -- -- | -- -- -- | -- -- -- | 45,000 48,000 60,000 | 25,000 30,000 35,000 | | | |
| | | Double Seam Welded Straight Seam Electric Straight Seam Submerged-Arc & Gas Metal-Arc Straight Seam | Non-expanded, X42**** | 0 29 | 1 25 | -- | 0 04 | -- | 0 05 | -- | -- | -- | 60,000 | 42,000 | | | |
| | | Combination Gas Metal-Arc, SAW Spiral Weld Pipe | X46**** X52**** | 0 31 0 31 | 1 35 1 35 | -- -- | 0 04 0 04 | -- -- | 0 05 0 05 | -- -- | -- -- | -- -- | 63,000 66,000** 72,000*** | 46,000 52,000 | | | |
| | | Butt-Weld (Applicable to Grade A25 only) Cold Expansion(except butt weld, shall be either non-expanded or cold expanded) | Cold expanded, X42**** X46**** X52**** | 0 29 0 29 0 29 | 1 25 1 25 1 25 | -- -- -- | 0 04 0 04 0 04 | -- -- -- | 0 05 0 05 0 05 | -- 0 005** 0 005** | -- 0 002** 0 002** | -- 0 003** 0 003** | 60,000 63,000 66,000** 72,000*** 71,000** 75,000*** | 42,000 46,000 52,000 56,000 | | | |
| | | | Non-expanded or cold, X56 X60 | 0 26 0 26 | 1 35 1 35 | -- -- | 0 04 0 04 | -- -- | 0 05 0 05 | 0 005** 0 005** | 0 002** 0 002** | 0 003** 0 003** | 75,000** 78,000*** | 60,000 | | | |
| | | | Expanded X65 X70 | (By Agreement) (By Agreement) | | | | | | | | | 77,000** 80,000*** 82,000 | 65,000 70,000 | | | |
| | | Electric-Weld Butt-Weld only | Welded Grade A25, Class I Grade A25, Class II Non-expanded or cold | 0 21 0 21 | 0 60 0 60 | 0 30 0 30 | 0 045 0 08 | -- 0 045 | 0 06 0 06 | -- -- | -- -- | -- -- | 45,000 45,000 | 25,000 25,000 | | | |

## Table 9-1. Summaries of API Specifications 5L

| Edition | Date | Materials | Processes | Grades | C Max | Mn Max | Mn Min | P Max | P Min | S Max | Cb Min | V Min | Ti Min | Tensile Str. psi | Yield Str. psi | Elong. | Hydrostatic Test P=2 S t/D, S ≤2500 psig | Comments |
|---|---|---|---|---|---|---|---|---|---|---|---|---|---|---|---|---|---|---|
|  |  |  |  | Grade A (Expanded) | 0.21 | 0.90 | -- | 0.04 | -- | 0.05 | -- | -- | -- | 48,000 | 30,000 |  |  |  |
|  |  |  |  | Grade B | 0.26 | 1.15 | -- | 0.04 | -- | 0.05 | -- | -- | -- | 60,000 | 35,000 |  |  |  |
|  |  |  |  | X42**** | 0.28 | 1.25 | -- | 0.04 | -- | 0.05 | -- | -- | -- | 60,000 | 42,000 |  |  |  |
|  |  |  |  | Non-expanded, X46**** | 0.30 | 1.35 | -- | 0.04 | -- | 0.05 | -- | -- | -- | 63,000 | 46,000 |  |  |  |
|  |  |  |  | X52**** | 0.30 | 1.35 | -- | 0.04 | -- | 0.05 | -- | -- | -- | 66,000****/72,000**** | 52,000 |  |  |  |
|  |  |  |  | Cold expanded, X46**** | 0.28 | 1.25 | -- | 0.04 | -- | 0.05 | -- | -- | -- | 66,000****/72,000**** | 46,000 |  |  |  |
|  |  |  |  | X52**** | 0.28 | 1.25 | -- | 0.04 | -- | 0.05 | -- | -- | -- | 72,000**** | 52,000 |  |  |  |
|  |  |  |  | Non-expanded or cold, X56 | 0.26 | 1.35 | -- | 0.04 | -- | 0.05 | 0.005* | 0.002* | 0.003* | 71,000****/75,000**** | 56,000 |  |  |  |
|  |  |  |  | X60 | 0.26 | 1.35 | -- | 0.04 | -- | 0.05 | 0.005* | 0.002* | 0.003* | 75,000****/78,000**** | 60,000 |  |  |  |
|  |  |  |  | Expanded, X65 | 0.026 | -- | 1.40 | -- | 0.04 | 0.05 | 0.005* | 0.002* | -- | 77,000****/80,000**** | 65,000 |  |  |  |
|  |  |  |  | X70 | 0.23 | 1.60 | -- | 0.04 | -- | 0.05 | -- | -- | -- | 82,000 | 70,000 |  |  |  |
| 34th API Spec 5L | May-84 | Same | Same | Same |  |  |  |  | Same |  |  |  |  | Same | Same |  | Same |  |
| 35th API Spec 5L | May-85 | Same | Same | Seamless |  |  |  |  |  |  |  |  |  | Same | Same |  | Same |  |
|  |  |  |  | Non-expanded or cold |  |  |  |  |  |  |  |  |  |  |  |  |  |  |
|  |  |  |  | Grade A25, Class I | 0.21 | 0.60 | 0.30 | 0.045 | -- | 0.06 | -- | -- | -- |  |  |  |  |  |
|  |  |  |  | Expanded |  |  |  |  |  |  |  |  |  |  |  |  |  |  |
|  |  |  |  | Grade A25, Class II** | 0.21 | 0.60 | 0.30 | 0.08 | 0.045 | 0.06 | -- | -- | -- |  |  |  |  |  |
|  |  |  |  | Grade A | 0.22 | 0.90 | -- | 0.04 | -- | 0.05 | -- | -- | -- |  |  |  |  |  |
|  |  |  |  | Grade B | 0.27 | 1.15 | -- | 0.04 | -- | 0.05 | -- | -- | -- |  |  |  |  |  |
|  |  |  |  | Non-expanded, X42**** | 0.29 | 1.25 | -- | 0.04 | -- | 0.05 | -- | -- | -- |  |  |  |  |  |
|  |  |  |  | X46**** | 0.31 | 1.35 | -- | 0.04 | -- | 0.05 | -- | -- | -- |  |  |  |  |  |
|  |  |  |  | X52**** | 0.31 | 1.35 | -- | 0.04 | -- | 0.05 | -- | -- | -- |  |  |  |  |  |
|  |  |  |  | Cold expanded, X42**** | 0.29 | 1.25 | -- | 0.04 | -- | 0.05 | -- | -- | -- |  |  |  |  |  |
|  |  |  |  | X46**** | 0.29 | 1.25 | -- | 0.04 | -- | 0.05 | -- | -- | -- |  |  |  |  |  |
|  |  |  |  | X52**** | 0.29 | 1.25 | -- | 0.04 | -- | 0.05 | -- | -- | -- |  |  |  |  |  |
|  |  |  |  | Non-expanded or cold, X56 | 0.26 | 1.35 | -- | 0.04 | -- | 0.05 | 0.005* | 0.002* | 0.003***** |  |  |  |  |  |
|  |  |  |  | X60 | 0.26 | 1.35 | -- | 0.04 | -- | 0.05 | 0.005* | 0.002* | 0.003***** |  |  |  |  |  |
|  |  |  |  | Expanded X65 | (By Agreement) |  |  |  |  |  |  |  |  |  |  |  |  |  |
|  |  |  |  | X70 | (By Agreement) |  |  |  |  |  |  |  |  |  |  |  |  |  |
|  |  |  |  | X80 | (By Agreement) |  |  |  |  |  |  |  |  | 90,000 | 80,000 |  |  |  |
|  |  |  |  | Welded |  |  |  |  |  |  |  |  |  |  |  |  |  |  |
|  |  |  |  | Electric-Weld |  |  |  |  |  |  |  |  |  |  |  |  |  |  |
|  |  |  |  | Grade A25, Class I | 0.21 | 0.60 | 0.30 | 0.045 | -- | 0.06 | -- | -- | -- |  |  |  |  |  |
|  |  |  |  | Butt-Weld only |  |  |  |  |  |  |  |  |  |  |  |  |  |  |
|  |  |  |  | Grade A25, Class II | 0.21 | 0.60 | 0.30 | 0.08 | 0.045 | 0.06 | -- | -- | -- |  |  |  |  |  |
|  |  |  |  | Non-expanded or cold |  |  |  |  |  |  |  |  |  |  |  |  |  |  |
|  |  |  |  | Grade A | 0.21 | 0.90 | -- | 0.04 | -- | 0.05 | -- | -- | -- |  |  |  |  |  |
|  |  |  |  | Expanded |  |  |  |  |  |  |  |  |  |  |  |  |  |  |
|  |  |  |  | Grade B | 0.26 | 1.15 | -- | 0.04 | -- | 0.05 | -- | -- | -- |  |  |  |  |  |
|  |  |  |  | X42**** | 0.28 | 1.25 | -- | 0.04 | -- | 0.05 | -- | -- | -- |  |  |  |  |  |
|  |  |  |  | Non-expanded, X46**** | 0.30 | 1.35 | -- | 0.04 | -- | 0.05 | -- | -- | -- |  |  |  |  |  |
|  |  |  |  | X52**** | 0.30 | 1.35 | -- | 0.04 | -- | 0.05 | -- | -- | -- |  |  |  |  |  |
|  |  |  |  | Cold expanded, X46**** | 0.28 | 1.25 | -- | 0.04 | -- | 0.05 | -- | -- | -- |  |  |  |  |  |
|  |  |  |  | X52**** | 0.28 | 1.25 | -- | 0.04 | -- | 0.05 | -- | -- | -- |  |  |  |  |  |
|  |  |  |  | Non expanded or cold, X56 | 0.26 | 1.35 | -- | 0.04 | -- | 0.05 | 0.005* | 0.002* | 0.003***** |  |  |  |  |  |

## Table 9-1. Summaries of API Specifications 5L

| Edition | Date | Materials | Processes | Grades | C Max | Mn Max | Mn Min | P Max | P Min | S Max | Cb Min | V Min | Ti Min | Tensile Str. psi | Yield Str. psi | Elong. | Hydrostatic Test P = 2 S t/D | S ≤2500 psig | Comments |
|---|---|---|---|---|---|---|---|---|---|---|---|---|---|---|---|---|---|---|---|
| | | | | X60 | 0.26 | 1.35 | -- | 0.04 | | 0.05 | 0.005* | 0.002* | 0.003**** | | | | | | |
| | | | | Expanded X65 | 0.26 | 1.40 | -- | | 0.04 | 0.05 | 0.005** | 0.002** | -- | | | | | | |
| | | | | X70 | 0.23 | 1.60 | -- | 0.04 | -- | 0.05 | -- | -- | -- | | | | | | |
| | | | | Non-expanded or Cold expanded X80 | 0.18 | 1.80 | -- | 0.03 | Same | 0.018 | -- | -- | -- | 90,000 | 80,000 | | | | |
| 36th Jun 1987 API Spec 5L | | Same | Same | Same | | | | | Same | | | | | | Same | | Same | | |
| 37th May-88 API Spec 5L | | Same | Same | Same | | | | | Same | | | | | | Same | | Same | | |
| 38th May-90 API Spec 5L | | Same | Seamless Without Filler Metal | Seamless Non-expanded or cold | | | | | | | | | | | Same | | Same | | |
| | | | Continuous Welding | Grade A25, Class I | 0.21 | 0.60 | 0.30 | 0.045 | -- | 0.06 | -- | -- | -- | | | | | | |
| | | | Electric-Welding With Filler Metal | Grade A25, Class II** Expanded | 0.21 | 0.60 | 0.30 | 0.08 | 0.045 | 0.06 | -- | -- | -- | | | | | | |
| | | | Submergd-Arc Arc Welding | Grade A | 0.22 | 0.90 | -- | 0.04 | -- | 0.05 | -- | -- | -- | | | | | | |
| | | | Gas Metal-Arc Welding | Grade B | 0.27 | 1.15 | -- | 0.04 | -- | 0.05 | -- | -- | -- | | | | | | |
| | | | Types of Pipe Seamless Welded Continuous | Non expanded, X42**** | 0.29 | 1.25 | -- | 0.04 | -- | 0.05 | -- | -- | -- | | | | | | |
| | | | | X46**** | 0.31 | 1.35 | -- | 0.04 | -- | 0.05 | -- | -- | -- | | | | | | |
| | | | | X52**** | 0.31 | 1.35 | -- | 0.04 | -- | 0.05 | -- | -- | -- | | | | | | |
| | | | Electric Longitudinal Seam | Cold expanded, X42**** | 0.29 | 1.25 | -- | 0.04 | -- | 0.05 | -- | -- | -- | | | | | | |
| | | | Submerged-Arc Gas Metal-Arc Combination Gas Metal-Arc and Submerged-Arc | X46**** | 0.29 | 1.25 | -- | 0.04 | -- | 0.05 | -- | -- | -- | | | | | | |
| | | | | X52**** | 0.29 | 1.25 | -- | 0.04 | -- | 0.05 | -- | -- | -- | | | | | | |
| | | | Double Seam Submerged-Arc Double Seam Gas metal-Arc | Non-expanded or cold, X56 | 0.26 | 1.35 | -- | 0.04 | -- | 0.05 | 0.005* | 0.005* | 0.005* | | | | | | |
| | | | Double Seam Combination Gas metal-Arc and Submerged-Arc Helical Seam Submerged-Arc | X60 | 0.26 | 1.35 | -- | 0.04 | -- | 0.05 | 0.005* | 0.005* | 0.005* | | | | | | |
| | | | Types of Seam Welds Electric Submerged-arc Gas Metal-Arc Skelp End Jointer | Expanded X65 | | | | (By Agreement) | | | | | | | | | | | |
| | | | | X70 | | | | (By Agreement) | | | | | | | | | | | |
| | | | | X80 | | | | (By Agreement) | | | | | | | | | | | |
| | | | Welded Tack | Electric-Weld Grade A25, Class II** | 0.21 | 0.60 | 0.30 | 0.045 | -- | 0.06 | -- | -- | -- | | | | | | |
| | | | Cold Expansion(except continuous welded, shall be either nonexpanded or cold expanded) | Continuous only Grade A25, Class II** | | | | | | | | | | | | | | | |
| | | | | Non-expanded or cold A | 0.21 | 0.90 | -- | 0.04 | -- | 0.05 | -- | -- | -- | | | | | | |
| | | | | Expanded Grade B | 0.26 | 1.15 | -- | 0.04 | -- | 0.05 | -- | -- | -- | | | | | | |
| | | | | X42**** | 0.28 | 1.25 | -- | 0.04 | -- | 0.05 | -- | -- | -- | | | | | | |
| | | | | Non-expanded, X46**** | 0.30 | 1.35 | -- | 0.04 | -- | 0.05 | -- | -- | -- | | | | | | |

## Table 9-1. Summaries of API Specifications 5L

| Edition | Date | Materials | Processes | Grades | C Max | Mn Max | Mn Min | P Max | P Min | S Max | S Min | Cb Min | V Min | Ti Min | Tensile Str. psi | Yield Str. psi | Elong. | Hydrostatic Test P = 2 S t/D  S ≤2500 psig | | Comments |
|---|---|---|---|---|---|---|---|---|---|---|---|---|---|---|---|---|---|---|---|---|
| | | | | X52**** | 0.30 | 1.35 | -- | 0.04 | -- | 0.05 | | -- | -- | -- | | | | | | |
| | | | | Cold expanded, X46**** | 0.28 | 1.25 | -- | 0.04 | -- | 0.05 | | -- | -- | -- | | | | | | |
| | | | | X52**** | 0.28 | 1.25 | -- | 0.04 | -- | 0.05 | | -- | -- | -- | | | | | | |
| | | | | Non expanded or cold, X56 | 0.26 | 1.35 | -- | 0.04 | -- | 0.05 | | 0.005** | 0.005** | 0.005**** | | | | | | |
| | | | | X60 | 0.26 | 1.35 | -- | 0.04 | -- | 0.05 | | 0.005** | 0.005** | 0.005**** | | | | | | |
| | | | | Expanded, X65 | 0.26 | 1.40 | -- | -- | 0.04 | 0.05 | | 0.005** | 0.005** | 0.005**** | | | | | | |
| | | | | X70 | 0.23 | 1.60 | -- | 0.04 | -- | 0.05 | | -- | -- | -- | | | | | | |
| | | | | Non-expanded or Cold expanded X80 | 0.18 | 1.80 | -- | 0.03 | -- | 0.018 | | -- | -- | -- | | | | | | |
| 39th  Jun 1991  API Spec 5L | | Same | Same | Same | | | Same | | | Same | | | | | Same | | | See book | | |
| 40th  Nov 1992  API Spec 5L | | Same | Same | Same  Expanded, X65  X70  X80 | | | Same | | | Same | | | | | Same | | | See book | | |

* In case of "Grade C" pipe is to be joined by welding, the purchaser may wish to stipulate the carbon content by special agreement
** Rephosphorized
*** Pipe made from killed, deoxidized, basic-Bessemer steel shall conform to the chemical requirements for pipe made from electric-furnace or O H steel, or from killed, deoxidized, acid-Bessemer steel, as specified on the purchase order
# Tentative
## For pipe > 20"O D
### For pipe 20"O D and larger
#### Pipe made from killed, deoxidized, basic-bessemer steel shall conform to the chemical requirements given for pipe made from electric furnace or O H steel, or from killed, deoxidized, acid-bessemer steel, as specified by purchaser order
* Limited to Butt Welded only
++ Open-hearth and basic-oxygen steel is rephosphorized
++++ Spiral Weld is limmited to 4 1/2" O.D. and larger
..... Dobule seam pipe is limited to 16"O D and larger
(a) Table given for elongations based on wall thickness

## Table 9-2. Summaries of API Specification 5LX.

| Edition | Date | Materials | Processes | Grades | C Max | Mn Max | Mn Min | P Max | P Min | S Max | S Min | Cb Min | V Min | Ti Min | Tensile Str. psi | Yield Str. psi | Elong. | Hydrostatic Test P = 2 S t/D  S <3000 psig | Comments |
|---|---|---|---|---|---|---|---|---|---|---|---|---|---|---|---|---|---|---|---|
| 1st API STD 5LX | Feb 1948 | Open-Hearth  Electric-Furnace Steel  Killed, deoxidized, acid bessemer steel | Seamless or mill-welding in which cold preformed skelp is longitudinally welded)  Electric-Flash Welding Continuous Electric-Resistance Welding  Submerged-Arc Welding | Grade X42 | 0.30L 0.33c | 1.25L 1.28c | — — | 0.11L 0.115c | — — | 0.06L 0.065c | — — | — — | — — | — — | 60,000 | 42,000 | | 85% of SMYS | |
| | | | L Ladle Analysis  C Check Analysis | | | | | | | | | | | | | | | | |
| 2nd API STD 5LX | May 1949 | Same | Seamless or mill-welding in which cold preformed plate or skelp is longitudinally butt welded)  Electric-Flash Welding Continuous Electric-Resistance Welding  Submerged-Arc Welded | Same | 0.30L 0.33c | 1.25L 1.28c | — — | 0.045La 0.080Lb 0.111Ld 0.055Ca 0.09Cb 0.11Cd | — — | 0.06L 0.065c | — — | — — | — — | — — | Same | | | 85% of SMYS | |
| | | | L Ladle Analysis  C Check Analysis  a Open-hearth and Electric-Furnace Steel  b Rephosphorized open-hearts Steel  d Killed, deoxidized, acid-bessemer steel | | | | | | | | | | | | | | | | |
| 3rd API STD 5LX Supplement No 1 | Mar 1951  Jan 1952 | Same | Same | Same | | | | Same | | | | | | | Same | | | 85% of SMYS | |
| 4th API STD 5LX Supplement No 1 | Mar 1953  Feb 1954 | Open-Hearth Electric-Furnace Steel Bessemer Steel Killed, deoxidized, acid bessemer steel Killed, deoxidized, basic-bessemer steel | Same | Grade X42  X46 X52 | 0.30L 0.34c | 1.25L 1.30c | — — | 0.045La 0.080Lb 0.111Ld 0.055Ca 0.09Cb 0.11Cd | — — | 0.06L 0.065c | — — | — — | — — | — — | 60,000  63,000 66,000 | 42,000  46,000 52,000 | | 85% of SMYS | |
| | | | L Ladle Analysis  C Check Analysis  a Open-hearth and Electric-Furnace Steel  b Rephosphorized open-hearts Steel  d Killed, deoxidized, acid-bessemer steel | | | | | | | | | | | | | | | | |
| 5th API STD 5LX | Nov 1954 | Same | Seamless or mill welding Electric-furnace or open-hearth  Non-expanded Non-expanded Cold-expanded Killed, deoxidized, acid-bessemer  Non-expanded | X42 X46, X52 X42, X46, X52  X42 X46, X52 | 0.29 0.32 0.29  0.24 0.27 | 1.25 1.35 1.25  1.25 1.35 | — — — — — | 0.04 0.04 0.04  0.10 0.10 | — — — — — | 0.05 0.05 0.05  0.05 0.05 | — — — — — | — — — — — | — — — — — | — — — — — | Same | | | 85% of SMYS | |

## Table 9-2. Summaries of API Specification 5LX.

| Edition | Date | Materials | Processes | Grades | C Max | Mn Max | Mn Min | P Max | P Min | S Max | Cb Min | V Min | Ti Min | Tensile Str. psi | Yield Str. psi | Elong. | P = 2 S t/D | S ≤3000 psig | Comments |
|---|---|---|---|---|---|---|---|---|---|---|---|---|---|---|---|---|---|---|---|
| | | | Cold expanded | X42, X46, X52 | 0.24 | 1.25 | -- | 0.10 | -- | 0.05 | -- | -- | -- | | | | | | |
| | | | Rephosphorized open-hearth | | | | | | | | | | | | | | | | |
| | | | Non-expanded | X42 | 0.29 | 1.25 | -- | 0.08 | -- | 0.05 | -- | -- | -- | | | | | | |
| | | | Cold-expanded | X42 | 0.29 | 1.25 | -- | 0.08 | -- | 0.05 | -- | -- | -- | | | | | | |
| | | | Welded | | | | | | | | | | | | | | | | |
| | | | Electric-flash | | | | | | | | | | | | | | | | |
| | | | Continuous electric-resistance | | | | | | | | | | | | | | | | |
| | | | Automatic submerged-arc using at least 2 weld passes | | | | | | | | | | | | | | | | |
| | | | Electric Welded | | | | | | | | | | | | | | | | |
| | | | Electric-furnace or open-hearth#### | | | | | | | | | | | | | | | | |
| | | | Non-expanded | X42 | 0.28 | 1.25 | -- | 0.04 | -- | 0.05 | -- | -- | -- | | | | | | |
| | | | Non-expanded | X46, X52 | 0.31 | 1.35 | -- | 0.04 | -- | 0.05 | -- | -- | -- | | | | | | |
| | | | Cold-expanded | X42, X46, X52 | 0.28 | 1.25 | -- | 0.04 | -- | 0.05 | -- | -- | -- | | | | | | |
| | | | Killed, deoxidized, acid-bessemer#### | | | | | | | | | | | | | | | | |
| | | | Non-expanded | X42 | 0.23 | 1.25 | -- | 0.10 | -- | 0.05 | -- | -- | -- | | | | | | |
| | | | Non-expanded | X46, X52 | 0.26 | 1.35 | -- | 0.10 | -- | 0.05 | -- | -- | -- | | | | | | |
| | | | Cold-expanded | X42, X46 | 0.23 | 1.25 | -- | 0.10 | -- | 0.05 | -- | -- | -- | | | | | | |
| | | | Cold-expanded | X52 | 0.24 | 1.25 | -- | 0.10 | -- | 0.05 | -- | -- | -- | | | | | | |
| | | | Rephosphorized open-hearth | | | | | | | | | | | | | | | | |
| | | | Non-expanded | X42 | 0.28 | 1.25 | -- | 0.08 | -- | 0.05 | -- | -- | -- | | | | | | |
| | | | Cold-expanded | X42 | 0.28 | 1.25 | -- | 0.08 | -- | 0.05 | -- | -- | -- | | | | | | |
| 6th | Feb 1956 | Same | Seamless or mill welding | | | | | | | | | | | | Same | | 85% of SMYS | | |
| API STD 5LX | | | Electric-furnace, open hearth,or killed, deoxidized, basic-bessemer#### | | | | | | | | | | | | | | | | |
| | | | Non-expanded | X42 | 0.29 | 1.25 | -- | 0.04 | -- | 0.05 | -- | -- | -- | | | | | | |
| | | | Non-expanded | X46, X52 | 0.32 | 1.35 | -- | 0.04 | -- | 0.05 | -- | -- | -- | | | | | | |
| | | | Cold-expanded | X42, X46, X52 | 0.29 | 1.25 | -- | 0.04 | -- | 0.05 | -- | -- | -- | | | | | | |
| | | | Killed, deoxidized, acid-bessemer, or killed, deoxidized, basic-bessemer#### | | | | | | | | | | | | | | | | |
| | | | Non-expanded | X42 | 0.24 | 1.25 | -- | 0.10 | -- | 0.05 | -- | -- | -- | | | | | | |
| | | | Non-expanded | X46, X52 | 0.27 | 1.35 | -- | 0.10 | -- | 0.05 | -- | -- | -- | | | | | | |
| | | | Cold-expanded | X42, X46, X52 | 0.24 | 1.25 | -- | 0.10 | -- | 0.05 | -- | -- | -- | | | | | | |
| | | | Welded | | | | | | | | | | | | | | | | |
| | | | Electric-flash | | | | | | | | | | | | | | | | |
| | | | Continuous electric-resistance | | | | | | | | | | | | | | | | |
| | | | Automatic submerged-arc using at least 2 weld passes | | | | | | | | | | | | | | | | |
| | | | Electric-furnace, open hearth, or killed, deoxidized, basic-bessemer#### | | | | | | | | | | | | | | | | |
| | | | Non-expanded | X42 | 0.28 | 1.25 | -- | 0.04 | -- | 0.05 | -- | -- | -- | | | | | | |
| | | | Non-expanded | X46, X52 | 0.31 | 1.35 | -- | 0.04 | -- | 0.05 | -- | -- | -- | | | | | | |
| | | | Cold-expanded | X42, X46, X52 | 0.28 | 1.25 | -- | 0.04 | -- | 0.05 | -- | -- | -- | | | | | | |

## Table 9-2. Summaries of API Specification 5LX.

| Edition Date | Materials | Processes | Grades | C Max | Mn Max | Mn Min | P Max | P Min | S Max | S Min | Cb Min | V Min | Ti Min | Tensile Str. psi | Yield Str. psi | Elong. | P = 2 S t/D, S <3000 psig (Hydrostatic Test) | Comments |
|---|---|---|---|---|---|---|---|---|---|---|---|---|---|---|---|---|---|---|
| | Killed, deoxidized, acid-bessemer, or killed, deoxidized, basic-bessemer#### | Non-expanded | X42 | 0.23 | 1.25 | -- | 0.10 | -- | 0.05 | -- | -- | -- | -- | | | | | |
| | | Non-expanded | X46, X52 | 0.26 | 1.35 | -- | 0.10 | -- | 0.05 | -- | -- | -- | -- | | | | | |
| | | Cold-expanded | X42, X46 | 0.23 | 1.25 | -- | 0.10 | -- | 0.05 | -- | -- | -- | -- | | | | | |
| | | Cold-expanded | X52 | 0.24 | 1.25 | -- | 0.10 | -- | 0.05 | -- | -- | -- | -- | | | | | |
| 7th Apr 1957 API STD 5LX | Same | Same | Same | Same | | | | | | | | | | Same | | 85% of SMYS | |
| 8th Mar 1958 API STD 5LX Supplement No 1 Mar 1959 | Same | Same | Same | Same | | | | | | | | | | Same | | 85% of SMYS | |
| atic tests, dimensions and manufacturers | | | | | | | | | | | | | | | | | |
| 9th Feb 1960 | Same | Seamless or mill welding(wrought tubular) | Same | Same | | | | | | | | | | Same | | 85% of SMYS | |
| API STD 5LX Supplement No 1 | | Same | | | | | | | | | | | | | | | |
| | | Welded(made by longitudinally butt welding cold-performed plate or skelp) | | | | | | | | | | | | | | | |
| Jan 1961 replaces process & manufacturers | Same | Same | | | | | | | | | | | | | | | |
| 10th Mar 1962 API STD 5LX Supplement No 1 Jul 1962 | Same | Same | Same | Same | | | | | | | | | | Same | | 85% of SMYS | |
| 11th Mar 1963 API STD 5LX Supplement No 1 Mar 1964 | Same | Same | Same | Same | | | | | | | | | | Same | | 85% of SMYS | |
| 12th Mar 1965 API STD 5LX | Same | Same | Same | Same | | | | | | | | | | Same | | 85% SMYS | |
| 13th Mar 1966 API STD 5LX | Same | Seamless or mill weldi Electric-furnace, open hearth, or killed, deoxidized, basic-bessemer#### | Same | Same | | | | | | | | | | | Same | | | |
| | | Non-expanded | | | | | | | | | | | | | | | |
| | | Non-expanded | | | | | | | | | | | | | | | |
| | | Cold-expanded | | | | | | | | | | | | | | | |
| | Killed, deoxidized, acid-bessemer, or killed, deoxidized, basic-bessemer#### | Non- or cold-expanded | X60 | 0.26 | 1.35 | -- | 0.04 | -- | 0.05 | -- | 0.05 | 0.01 | 0.02 | -- | 75,000 | 60,000 | | | |
| | | Non-expanded | | | | | | | | | | | | | | | |
| | | Non-expanded | | | | | | | | | | | | | | Same | | |
| | | Cold-expanded | | | | | | Same | | | | | | | | Same | | |
| | Welded Electric-flash Electric-resistance* Automatic submerged-arc using at least 2 weld passes | | Same | | | | | Same | | | | | | | | Same | | |
| | Electric-furnace, open-hearth, or killed, deoxidized, basic-bessemer#### Non-expanded | | | | | | | | | | | | | | | | 85% SMYS | |

## Table 9-2. Summaries of API Specification 5LX.

| Edition Date | Materials | Processes | Grades | C Max | Mn Max | Mn Min | P Max | P Min | S Max | Cb Min | V Min | Ti Min | Tensile Str. psi | Yield Str. psi | Elong. | Hydrostatic Test P = 2 S t/D, S <3000 psig | Comments |
|---|---|---|---|---|---|---|---|---|---|---|---|---|---|---|---|---|---|
| | | Non-expanded Cold-expanded | X60 | 0.26 | 1.35 | -- | 0.04 | -- | 0.05 | 0.01 | 0.02 | -- | 75,000 | 60,000 | | | |
| | | Non- or cold-expanded Killed, deoxidized, acid-bessemer, or killed, deoxidized, basic-bessemer##### Non-expanded Non- or cold-expanded Cold-expanded Cold-expanded | Same | | | | Same | | | | | | Same | | | 85% SMYS | |
| | | * Not permissible for the manufacture of Grade X60 pipe except by agreement | | | | | | | | | | | | | | | |
| 14th Mar 1967 API STD 5LX | Same | Seamless  Same | Same | | | | Same | | | | | | Same | | | 85% SMYS | |
| | | | X52 | | 1.35 | | | | | | | | 66,000-72, | 52,000 | | | |
| | | Non- or cold-expanded | X60 | 0.26 | 1.35 | -- | 0.04 | -- | 0.05 | 0.005 | 0.02 | | 75,000-78, | 60,000 | | | |
| | | Non- or cold-expanded | X65 | | | | (By Agreement) | | | | | 0.03 | 80,000 | 65,000 | | | |
| | | Welded  Same | Same | | | | Same | | | | | | Same | | | | |
| | | Non- or cold-expanded | X60 | 0.26 | 1.35 | -- | 0.04 | -- | 0.05 | 0.005 | 0.02 | 0.03 | 75,000-78, | 60,000 | | | |
| | | Non- or cold-expanded | X65 | 0.26 | 1.40 | -- | 0.04 | -- | 0.05 | 0.005 | 0.02 | -- | 80,000 | 65,000 | | | |
| | | Same | Same | | | | Same | | | | | | Same | | | | |
| 15th Mar 1968 API STD 5LX | Same | Seamless  Same | Same | | | | Same | | | | | | Same | | | 85% SMYS | |
| | | | X52 | | 1.35 | | | | | | | | 66,000-72, | 52,000 | | | |
| | | Non- or cold-expanded | X56 X60 | 0.26 | 1.35 | -- | 0.04 | -- | 0.05 | 0.005 | 0.02 | | 75,000-78, | 60,000 | | | |
| | | Non- or cold-expanded | X65 | | | | (By Agreement) | | | | | 0.03 | 80,000 | 65,000 | | | |
| | | Welded  Same | Same | | | | Same | | | | | | Same | | | | |
| | | Non- or cold-expanded | X56 X60 | 0.26 | 1.35 | -- | 0.04 | -- | 0.05 | 0.005 | 0.02 | 0.03 | 75,000-78, | 60,000 | | | |
| | | Non- or cold-expanded | X65 | 0.26 | 1.40 | -- | 0.04 | -- | 0.05 | 0.005 | 0.02 | -- | 80,000 | 65,000 | | | |
| | | Same | Same | | | | Same | | | | | | Same | | | | |
| 16th Apr 1969 API STD 5LX | Same | Seamless Electric-weld Submerged-arc Weld Cold Expansion | Same X56 X60 X65 | | | | Same | | | | | | 75,000-71, 78,000-75, 80,000-77, | 56,000 60,000 65,000 | | 85% SMYS | |
| 17th Apr 1970 API STD 5LX | Same | Same | Seamless | | | | | | | | | | Same | | | 85% SMYS | |
| | | | Non-expanded, X42 | 0.29 | 1.25 | -- | 0.04 | -- | 0.05 | -- | -- | -- | | | | | |
| | | | Non-expanded, X46,X52 | 0.31 | 1.35 | -- | 0.04 | -- | 0.05 | -- | -- | -- | | | | | |
| | | | Cold-expanded X42, X46, X52 | 0.29 | 1.25 | -- | 0.04 | -- | 0.05 | -- | -- | -- | | | | | |
| | | | Non- or cold-expanded X56, X60 | 0.26 | 1.35 | -- | 0.04 | -- | 0.05 | 0.005 | 0.02 | 0.03 | | | | | |
| | | | Non- or cold-expanded X65 | | | | (By Agreement) | | | | | | | | | | |
| | | | Welded Non-expanded, X42 | 0.28 | 1.25 | -- | 0.04 | -- | 0.05 | -- | -- | -- | | | | | |
| | | | Non-expanded, X46,X52 | 0.30 | 1.35 | -- | 0.04 | -- | 0.05 | -- | -- | -- | | | | | |

## Table 9-2. Summaries of API Specification 5LX.

| Edition Date | Materials | Processes | Grades | C Max | Mn Max | Mn Min | P Max | P Min | S Max | Cb Min | V Min | Ti Min | Tensile Str. psi | Yield Str. psi | Elong. | Hydrostatic Test P = 2 S t/D, S <3000 psig | Comments |
|---|---|---|---|---|---|---|---|---|---|---|---|---|---|---|---|---|---|
| 18th Apr 1971 API STD 5LX Supplement No 1 Apr 1972 | | | Cold-expanded X42, X46, X52 | 0.28 | 1.25 | -- | 0.04 | -- | 0.05 | -- | -- | -- | | | | | |
| | | | Non- or cold-expanded X56, X60 | 0.26 | 1.35 | -- | 0.04 | -- | 0.05 | 0.005 | 0.02 | -- | | | | | |
| | | | Non- or cold-expanded X65 | 0.26 | 1.40 | -- | 0.04 | -- | 0.05 | 0.005 | 0.02 | 0.03 | | | | | |
| | Same | Same | Same | | | | Same | | | | | | | Same | | 85% SMYS | |
| 19th Mar 1973 API Spec 5LX Supplement No 1 Mar 1972 *(changes to process, & manufacturers)* | Same | Seamless Electric-weld Submerged-arc Weld Submerged-arc Weld, Double Seam Cold Expansion | Seamless Non-expanded, X42 Non-expanded, X46,X52 Cold-expanded X42, X46, X52 Non- or cold-expanded X56, X60 Non- or cold-expanded X65, X70 Welded Non-expanded, X42 Non-expanded, X46,X52 Cold-expanded X42, X46, X52 Non- or cold-expanded X56, X60 Non- or cold-expanded X65 | | | | Same | | | | | | | Same | | 85% SMYS | |
| | | | X70 | 0.23 | 1.60 | -- | 0.04 | -- | 0.05 | -- | -- | -- | 82,000 | 70,000 | | | |
| 20th Mar 1975 API Spec 5LX Supplement No 1 Mar 1976 | Same | Same | Same | | | | Same | | | | | | | Same | | 85% SMYS | |
| 21st Mar 1977 API Spec 5LX | Same | Same | Same | | | | Same | | | | | | | Same | | 85% SMYS | |
| 22nd Mar 1978 API Spec 5LX | Same | Seamless Electric Weld submerged-arc Weld Gas Metal-arc Weld (MIG) Double Seam Welded Pipe Cold Expansion | Same | | | | Same | | | | | | | Same | | 85% SMYS | |
| 23rd Mar 1980 API Spec 5LX Supplement No 1 Mar 1981 | Same | Seamless Electric Weld Submerged-arc Weld Gas Metal-arc Weld (MIG) Combination Gas Metal-arc Weld (MIG) and Submerged Arc Weld Double Seam Welded Pipe Cold Expansion | Same | | | | Same | | | | | | Same | | | 85% SMYS | |

## Table 9-2. Summaries of API Specification 5LX.

| Edition Date | Materials | Processes | Grades | Chemical Properties | | | | | | | | | Tensile Properties | | | Hydrostatic Test | Comments |
|---|---|---|---|---|---|---|---|---|---|---|---|---|---|---|---|---|---|
| | | | | C Max | Mn Max | Mn Min | P Max | P Min | S Max | Cb Min | V Min | Ti Min | Tensile Str. psi | Yield Str. psi | Elong. | $P = 2\,S\,t/D$ | |
| 24th Mar. 1982 API Spec 5LX | Same | Seamless  Electric Weld  Submerged-arc Weld  Gas Metal-arc Weld (MIG)  Combination Gas Metal-arc Weld (MIG) and Submerged Arc Weld  Double Seam Welded (Applicable to sizes 36" O D and larger)  Cold Expansion | Same | Same | | | | | | | | | Same | | | 85% SMYS    S ≤3000 psig. | |

* In case of "Grade C" pipe is to be joined by welding, the purchaser may wish to stipulate the carbon content by special agreement
** Rephosphorized
*** Pipe made from killed, deoxidized, basic-Bessemer steel shall conform to the chemical requirements for pipe made from electric-furnace or O H steel, or from killed, deoxidized, acid-Bessemer steel, as specified on the purchase order
# Tentative
## For pipe > 20"O D
### For pipe 20"O D and larger
#### Pipe made from killed, deoxidized, basic-bessemer steel shall conform to the chemical requirements given for pipe made from electric furnace or O H steel, or from killed, deoxidized, acid-bessemer steel, as specified by purchaser order
+ Limited to Butt Welded only
++ Open-hearth and basic-oxygen steel is rephosphorized
+++ Spiral Weld is limited to 42 O D and larger
++++ Double seam pipe is limited to 36"O D and larger
(a) Table given for elongations based on wall thickness

## Table 9-3. North American Manufacturers Who Held 5L Licenses

| Manufacturers of API Line Pipe / 5L | 3rd Ed Jan-1930 | 4th Ed Jul-1931 | 5th Ed Jan-1934 | 6th Ed Aug-1935 | 7th Ed Apr-1940 | 8th Ed May-1942 | 9th Ed Aug-1944 | 10th Ed Aug-1945 | 11th Ed May-1949 | 12th Ed Mar-1951 | 13th Ed Mar-1954 | 14th Ed Mar-1955 | 15th Ed Mar-1956 | 16th Ed Apr-1957 | 17th Ed Mar-1958 | 18th Ed Feb-1960 | 19th Ed Mar-1962 |
|---|---|---|---|---|---|---|---|---|---|---|---|---|---|---|---|---|---|
| Aceros Alfa Monterrey, SA., Monterrey, N.L., Mexico | | | | | | | | | | | | | | | | | x |
| Acme-Newport Steel Co., New Port, KY | | | | | | | | | | | | | | x | x | x | x |
| Acme-Newport Steel Co., New Port, KY(Div of Acme Steel Co., Newport, KY(P-E only) | | | | | | | | | | | | | | | | | x |
| Alberta Phoenix Tube & Pipe Ltd., Edmonton, Alberta, Canada | | | | | | | | | | | | | | x | x | x | x |
| Algoma Steel Corp., Ltd., Sault Ste. Marie, Canada | | | | | | | | | | | | | | | | | |
| Allegheny Steel Company, Brackenridge, PA | x | | x | x | x | x | x | x | | | | | | | | | |
| American Bridge Div, Pittsburgh, PA (United States Steel) | | | | | | | | | | | | | | | | | |
| American Pipe & Construction Co, Portland, OR | | | | | | | | | | | | | | | | | |
| American Pipe Inspection Inc., Houston, TX | | | | | | | | | | | | | | | | | |
| American Rolling Mill Company, Middletown, OH | | | | | x | x | x | x | | | | | | | | | |
| American Seamless Tube Corporation, Los Angeles, CA | x | x | | | | | | | | | | | | | | | |
| American Steel Export Company, Inc., New York, NY | x | x | | | | | | | | | | | | | | | |
| American Steel Pipe, Birmingham, AL | | | | | | | | | | | | | | | | | |
| AMERON, Portland, OR | | | | | | | | | | | | | | | | | |
| Armco Div., Armco Steel Corp., Ambridge, PA | | | | | | | | | | | | | | | | | x |
| Armco Inc., Houston, TX | | | | | | | | | | | | | | | | | |
| Armco Steel Corp., Middletown, OH | | | | | | | | | | | | x | | | | | |
| Babcock & Wilcox Tube Co., Beaver Falls, PA | x | x | x | x | x | x | x | x | x | x | x | | | | | | |
| Beall Pipe and Tank Corp. Portland, OR | | | | | | | | | | | | | x | x | x | x | x |
| Berg Steel Pipe Corp., Panama City, FL | | | | | | | | | | | | | | | | | |
| Bethlehem Steel Co., Bethlehem, PA / Steelton, PA | x | x | x | x | x | x | x | x | x | x | x | x | x | x | x | x | x |
| Brooks Tube Ltd., Brooks, Alberta, Canada | | | | | | | | | | | | | | | | | |
| Bull Moose Tube Co., Gerald, MO | | | | | | | | | | | | | | | | | |
| Byers, A.M., Co., Pittsburgh, PA | x | x | x | x | x | x | x | x | x | x | x | x | x | x | x | x | x |
| Cal-Metal Pipe Corp. of Louisiana, Baton Rouge, LA | | | | | | | | | | | | | | | x | x | x |
| Cal-Metal Corp., Torrance, CA (P-E only) | | | | | | | | | | | | | | | x | x | |
| Cameron Iron Works, Inc., Houston, TX | | | | | | | | | | | | | | | | | |

## Table 9-3. North American Manufacturers Who Held 5L Licenses

| Manufacturers of API Line Pipe 5L | 20th Ed Mar-1963 | 21st Ed Mar-1965 | 22nd Ed Mar-1967 | 23rd Ed Mar-1968 | 24th Ed Apr-1969 | 25th Ed Apr-1970 | 26th Ed Apr-1971 | 27th Ed Mar-1973 | 28th Ed Mar-1975 | 29th Ed Mar-1977 | 30th Ed Mar-1978 | 31th Ed Mar-1980 | 32nd Ed Mar-1982 | 33rd Ed Mar-1983 | 34th Ed Mar-1984 | 35th Ed May-1985 | 36th Ed Jun-1987 |
|---|---|---|---|---|---|---|---|---|---|---|---|---|---|---|---|---|---|
| Aceros Alfa Monterrey, SA., Monterrey, N.L. Mexico | x | x | x | x | x | x | x | x | | | | | | | | | |
| Acme-Newport Steel Co., New Port, KY | x | | | | | | | | | | | | | | | | |
| Acme-Newport Steel Co., New Port, KY(Div of Acme Steel Co., Newport, KY[P-E only] | x | | | | | | | | | | | | | | | | |
| Alberta Phoenix Tube & Pipe Ltd., Edmonton, Alberta, Canada | x | x | x | | | | | | | | | | | | | | |
| Algoma Steel Corp., Ltd., Sault Ste. Marie, Canada | | | | | | | x | x | x | x | x | x | x | x | x | x | x |
| Allegheny Steel Company, Brackenridge, PA | | | | | | | | | | | | | | | | | |
| American Bridge Div, Pittsburgh, PA (United States Steel) | | x | x | x | x | x | x | x | x | | | | | | | | |
| American Pipe & Construction Co., Portland, OR | | x | x | x | x | x | | | | | | | | | | | |
| American Pipe Inspection Inc., Houston, TX | | | | | | | | | | | | | | | | x | |
| American Rolling Mill Company, Middletown, OH | | | | | | | | | | | | | | | | | |
| American Seamless Tube Corporation, Los Angeles, CA | | | | | | | | | | | | | | | | | |
| American Steel Export Company, Inc., New York, NY | | | | | | | | | | | | | | | | | |
| American Steel Pipe, Birmingham, AL | x | x | x | x | x | x | x | x | x | x | x | x | x | x | x | x | x |
| AMERON, Portland, OR | | | | | | | x | | | | | | | | | | |
| Armco Div. Armco Steel Corp., Ambridge, PA | x | x | x | x | x | x | x | x | | | | | | | | | |
| Armco Inc., Houston, TX | | | | | | | | | x | x | x | x | x | x | x | x | x |
| Armco Steel Corp. Middletown, OH | | | | | | | | | | | | x | x | | | | |
| Babcock & Wilcox Tube Co., Beaver Falls, PA | | | | | | | | | | | | | | | | | |
| Beall Pipe and Tank Corp. Portland, OR | x | x | x | x | x | x | x | x | x | x | x | x | x | | | | |
| Berg Steel Pipe Corp., Panama City, FL | | | | | | | | | | | | x | x | x | x | x | x |
| Bethlehem Steel Co., Bethlehem, PA / Steelton, PA | x | x | x | x | x | x | x | x | x | x | x | x | x | x | x | | |
| Brooks Tube Ltd., Brooks, Alberta, Canada | | | | | | | | | x | x | x | x | x | x | | | |
| Bull Moose Tube Co., Gerald, MO | | | | | | | | | | | | x | x | | | | |
| Byers, A.M., Co., Pittsburgh, PA | x | x | | | | | | | | | | | | | | | |
| Cal-Metal Pipe Corp. of Louisiana, Baton Rouge, LA | | | | | | | | | | | | | x | | | | |
| Cal-Metal Corp., Torrance, CA (P-E only) | x | x | x | x | x | x | x | x | x | x | x | x | x | | | | |
| Cameron Iron Works, Inc., Houston, TX | x | x | x | x | x | x | x | x | x | x | | | | x | x | x | x |

## Table 9-3. North American Manufacturers Who Held 5L Licenses

| Manufacturers of API Line Pipe / 5L | 3rd Ed Jan-1930 | 4th Ed Jul-1931 | 5th Ed Jan-1934 | 6th Ed Aug-1935 | 7th Ed Apr-1940 | 8th Ed May-1942 | 9th Ed Aug-1944 | 10th Ed Aug-1945 | 11th Ed May-1949 | 12th Ed Mar-1951 | 13th Ed Mar-1954 | 14th Ed Mar-1955 | 15th Ed Mar-1956 | 16th Ed Apr-1957 | 17th Ed Mar-1958 | 18th Ed Feb-1960 | 19th Ed Mar-1962 |
|---|---|---|---|---|---|---|---|---|---|---|---|---|---|---|---|---|---|
| Canadian Phoenix Steel & Pipe Ltd., Edmonton, Alberta, Canada | | | | | | | | | | | | | | | | | |
| Canadian Western Pipe Mills, Ltd., Port Moody, B.C., Canada | | | | | | | | | | | | | x | x | x | x | x |
| Central Steel Tube Co., Clinton, IO | | | | | | | | | | | | | | | | | |
| Central Tube Company, Pittsburgh, PA | x | x | x | x | | | | | | | | | | | | | |
| Chemetron Corp., Louisville, KY (Tube Turns) | | | | | | | | | | | | | | | | | |
| Colorado Fuel & Iron Corp., Pueblo, CO | | | | | | | | | | | | x | x | x | x | x | x |
| Consolidated Western Steel Corp., Los Angeles, CA | | | | | | | | | x | x | x | x | x | x | x | x | x |
| Copperweld Tubing Group, Shelby, OH — American Seamless Tubing Facility, Baltimore, MD | | | | | | | | | | | | | | | | | |
| Copperweld Tubing Group, Shelby, OH — Ohio Steel Tube Facility, Shelby, OH | | | | | | | | | | | | | | | | | |
| Copperweld Tubing Group, Shelby, OH — Regal Tube Facility, Chicago, IL | | | | | | | | | | | | | | | | | |
| Cyclops Corp. — Sawhill Tubular Div, Sharon, PA | | | | | | | | | | | | | | | | | |
| Cyclops Corp. — Wheatland, PA | | | | | | | | | | | | | | | | | |
| Cyclops Corp. — Tex-Tube Div | | | | | | | | | | | | | | | | | |
| Dominion Steel and Coal Corp., Montreal, Quebec, Canada | | | | | | | | | | | | | | | | x | x |
| Donovan Steel Tube Co., Toledo, OH | | | | | | | | | | | | | | | | | |
| Dosco Steel Ltd., Montreal, Quebec, Canada | | | | | | | | | | | | | | | | | |
| Fort Collins Pipe Co., Fort Collins, CO | | | | | | | | | | | | | | | | | |
| Fort Worth Pipe & Supply, Fort Worth, TX | | | | | | | | | | | | | | | | | |
| Fox Steel Pipe Corp., Jacksonville, FL(P-E only) | | | | | | | | | | | | | | | | | x |
| Geneva Tube, Geneva, NE | | | | | | | | | | | | | | | | | |
| Globe Steel Tubes Co., Milwaukee, WI | | | | x | x | x | x | x | x | x | x | x | x | | | | |
| Gulf + Western Mfg. Co., Oak Brook, IL | | | | | | | | | | | | | | | | | |
| Gulf States Tube Corp., Rosenberg, TX (P-E only) | | | | | | | | | | | | | | | | | |
| Harrisburg Steel Corp., Harrisburg, PA | | | | | x | x | x | x | x | | | | | | | | |

## Table 9-3. North American Manufacturers Who Held 5L Licenses

| Manufacturers of API Line Pipe 5L | 20th Ed Mar-1963 | 21st Ed Mar-1965 | 22nd Ed Mar-1967 | 23rd Ed Mar-1968 | 24th Ed Apr-1969 | 25th Ed Apr-1970 | 26th Ed Apr-1971 | 27th Ed Mar-1973 | 28th Ed Mar-1975 | 29th Ed Mar-1977 | 30th Ed Mar-1978 | 31st Ed Mar-1980 | 32nd Ed Mar-1982 | 33rd Ed Mar-1983 | 34th Ed Mar-1984 | 35th Ed May-1985 | 36th Ed Jun-1987 |
|---|---|---|---|---|---|---|---|---|---|---|---|---|---|---|---|---|---|
| Canadian Phoenix Steel & Pipe Ltd., Edmonton, Alberta, Canada | | | | x | x | x | x | x | | | | | | | | | |
| Canadian Western Pipe Mills, Ltd., Port Moody, B.C., Canada | x | x | x | | | | | | | | | | | | | | |
| Central Steel Tube Co., Clinton, IO | | | | | | | | | | | x | x | x | x | x | x | x |
| Central Tube Company, Pittsburgh, PA | | | | | | | | | | | | | | | | | |
| Chemetron Corp., Louisville, KY (Tube Turns) | | | | | | | x | x | x | x | x | x | x | x | | | |
| Colorado Fuel & Iron Corp., Pueblo, CO | x | x | x | x | x | x | x | x | x | x | x | x | x | x | x | x | x |
| Consolidated Western Steel Corp., Los Angeles, CA | x | | | | | | | | | | | | | | | | |
| Copperweld Tubing Group, Shelby, OH | | | | | | | | | | | | | | | | | |
|   American Seamless Tubing Facility, Baltimore, MD | | | | | | | | | | | | | | | x | | |
|   Ohio Steel Tube Facility, Shelby, OH | | | | | | | | | | | | | | | | x | x |
|   Regal Tube Facility, Chicago, IL | | | | | | | | | | | | | | | | x | x |
| Cyclops Corp. — Sawhill Tubular Div, Sharon, PA | | | | | x | x | x | x | x | x | x | x | x | x | x | x | x |
| Cyclops Corp. — Wheatland, PA | | | | | | | | | | | | | | | | x | |
| Cyclops Corp. — Tex-Tube Div | | | | | | | | | | | x | x | x | x | x | x | x |
| Dominion Steel and Coal Corp., Ltd., Montreal, Quebec, Canada | x | | | | | | | | | | | | | | | | |
| Donovan Steel Tube Co., Toledo, OH | | | | | | | | | | | | | x | x | | | |
| Dosco Steel Ltd., Montreal, Quebec, Canada | | x | x | x | x | x | x | x | | | | | | | | | |
| Fort Collins Pipe Co., Fort Collins, CO | | | | | | | | | | | | | | | | x | x |
| Fort Worth Pipe & Supply, Fort Worth, TX | | | | | | | | | | | | x | x | x | x | x | x |
| Fox Steel Pipe Corp., Jacksonville, FL (P-E only) | | | | | | | | | | | | | | | | | |
| Geneva Tube, Geneva, NE | | | | | | | | | | | | | | | x | x | x |
| Globe Steel Tubes Co., Milwaukee, WI | | | | | | | | | | | | | | | | | |
| Gulf + Western Mfg. Co., Oak Brook, IL | | | | | | | | | | x | x | | | | | | |
| Gulf States Tube Corp., Rosenberg, TX (P-E only) | | | | | | | | | | x | x | | | | | | |
| Harrisburg Steel Corp., Harrisburg, PA | x | x | x | x | x | x | x | x | x | x | | | | | | | |

## Table 9-3. North American Manufacturers Who Held 5L Licenses

| Manufacturers of API Line Pipe 5L | Location | 3rd Ed Jan-1930 | 4th Ed Jul-1931 | 5th Ed Jan-1934 | 6th Ed Aug-1935 | 7th Ed Apr-1940 | 8th Ed May-1942 | 9th Ed Aug-1944 | 10th Ed Aug-1945 | 11th Ed May-1949 | 12th Ed Mar-1951 | 13th Ed Mar-1954 | 14th Ed Mar-1955 | 15th Ed Mar-1956 | 16th Ed Apr-1957 | 17th Ed Mar-1958 | 18th Ed Feb-1960 | 19th Ed Mar-1962 |
|---|---|---|---|---|---|---|---|---|---|---|---|---|---|---|---|---|---|---|
| HYLSA, S.A., Monterrey, Mexico | | | | | | | | | | | | | | | | | | |
| Interlake Steel Corp., Newport, KY | | | | | | | | | | | | | | | | | | |
| International Portable Pipe Mills, Ltd., Calgary, Alberta, Canada | | | | | | | | | | | | | | | | | | |
| IPSCO, Regina, Canada | | | | | | | | | | | | | | | | | | x |
| | Calgary, Alberta, Canada Facility | | | | | | | | | | | | | | | | | |
| | Red Deer, Canada Facility | | | | | | | | | | | | | | | | | |
| | Port Moody, B.C., Canada Facility | | | | | | | | | | | | | | | | | |
| | Edmonton, Alberta, Canada | | | | | | | | | | | | | | | | | |
| Jones & Laughlin Steel Corporation, Aliquippa, PA | | | | | | | | | | | | | | | | | x | x |
| Jones & Laughlin Steel Corporation, Pittsburgh, PA | | x | x | x | x | x | x | x | x | x | x | x | x | x | x | x | | |
| Kaiser Pipe & Casing, Inc., Irwindale, CA | | | | | | | | | | | | | | | | | | |
| Kaiser Steel Corp., Fontana, CA | Napa, CA | | | | | | | | | | x | x | x | x | x | x | x | |
| | Oakland, CA | | | | | | | | | | | | | | | | | x |
| Kane Boiler Works, Inc., Galveston, TX | | | | | | | | | | | | | | | | | | |
| Kane, Industries, Inc., Galveston, TX | | | | | | | | | | | | | | | | | | |
| Laclede Steel Company, St. Louis, MO | Chicago, IL | | | | | | | | | | | | | | | | | |
| | South Chicago, Il | | | | | | | | | | | | | | | | | |
| | Cleveland, OH | | | | | | | | | | | | | | | | | |
| Lone Star Steel Co., Dallas, TX | Elyria, OH | | | | | | | | | | | x | x | x | x | x | x | x |
| Levine Suply Co., Houston, TX | Campbell, OH | | | | | | | | | | | | | | | | | |
| LTV Steel Corp., Youngstown, OH | Aliquippa, PA | | | | | | | | | | | | | | | | | |
| | Counce, TN | | | | | | | | | | | | | | | | | |
| Mannesmann Tube Co., Ltd., Sault Ste. Marie, Ont., Canada | | | | | | | | | | | | | | | | x | x | x |
| Maruichi American Corp., Santa Fe Springs, CA | | | | | | | | | | | | | | | | | | |
| Maverick Tube Corp., St. Louis, MO | | | | | | | | | | | | | | | | | | |
| Midwest Speciality, Tulsa, OK | | | | | | | | | | | | | | | | | | |

## Table 9-3. North American Manufacturers Who Held 5L Licenses

| Manufacturers of API Line Pipe 5L | 20th Ed Mar-1963 | 21st Ed Mar-1965 | 22nd Ed Mar-1967 | 23rd Ed Mar-1968 | 24th Ed Apr-1969 | 25th Ed Apr-1970 | 26th Ed Apr-1971 | 27th Ed Mar-1973 | 28th Ed Mar-1975 | 29th Ed Mar-1977 | 30th Ed Mar-1978 | 31st Ed Mar-1980 | 32nd Ed Mar-1982 | 33rd Ed Mar-1983 | 34th Ed Mar-1984 | 35th Ed May-1985 | 36th Ed Jun-1987 |
|---|---|---|---|---|---|---|---|---|---|---|---|---|---|---|---|---|---|
| HYLSA, S.A., Monterrey, Mexico | | | | | | | | | x | x | x | x | x | x | x | x | x |
| Interlake Steel Corp., Newport, KY | | x | x | x | x | x | x | x | x | x | x | | | | | x | |
| International Portable Pipe Mills, Ltd, Calgary, Alberta, Canada | | | | | | | | | | | | | | | | | |
| IPSCO, Regina, Canada | x | x | x | x | x | x | x | x | x | x | x | x | x | x | x | x | x |
|   Calgary, Alberta, Canada Facility | | | | | | | | | | | | | | | | x | x |
|   Red Deer, Canada Facility | | | | | | | | | | | | | | | | | x |
|   Port Moody, B.C., Canada Facility | | | | | | | | | | | | | | | | x | x |
|   Edmonton, Alberta, Canada | | | | | | | | | | | | | | | | | x |
| Jones & Laughlin Steel Corporation, Aliquippa, PA | x | x | x | x | x | x | x | x | x | | | | | | | | |
| Jones & Laughlin Steel Corporation, Pittsburgh, PA | | | | | | | | | | x | x | x | x | x | x | | x |
| Kaiser Pipe & Casing, Inc., Irwindale, CA | | | | | | | | | | | | | | x | x | x | x |
| Kaiser Steel Corp., Fontana, CA | | | | | | | | x | | x | x | x | x | x | x | x | x |
|   Napa, CA | | | | | | | | x | | | | | | x | | x | x |
|   Oakland, CA | x | x | x | x | x | x | x | x | x | | | | | | | | |
| Kane Boiler Works, Inc., Galveston, TX | | | | | | | x | x | | | | | | | | | |
| Kane, Industries, Inc., Galveston, TX | | | | | | | | | | | | | | | | | x |
| Laclede Steel Company, St. Louis, MO | x | x | x | x | x | x | x | x | x | x | x | x | x | x | x | x | x |
| Lone Star Steel Co., Dallas, TX | x | x | x | x | x | x | x | x | x | x | x | x | x | x | x | x | x |
| Levine Suply Co., Houston, TX | | | x | | | | | | | | | x(EF) | | | | | |
| LTV Steel Corp., Youngstown, OH | | | | | | | | | | | | | | | | x | x |
|   Chicago, IL | | | | | | | | | | | | | | | | x | x |
|   South Chicago, Il | | | | | | | | | | | | | | | | x | x |
|   Cleveland, OH | | | | | | | | | | | | | | | | x | x |
|   Elyria, OH | | | | | | | | | | | | | | | | x | x |
|   Campbell, OH | | | | | | | | | | | | | | | x | x | x |
|   Aliquippa, PA | | | | | | | | | | | | | | | x | x | x |
|   Counce, TN | | | | | | x | | | | | | | | | | | |
| Mannesmann Tube Co., Ltd., Sault Ste. Marie, Ont., Canada | x | x | x | x | x | x | | x | x | | | | | | | x | |
| Maruichi American Corp., Santa Fe Springs, CA | | | | | | | | | | | | x | | x | | x | x |
| Maverick Tube Corp., St. Louis, MO | | | | | | | | | | | x | x | x | x | x | x | x |
| Midwest Speciality, Tulsa, OK | | | | | | | | | | | | | | | | | x |

## Table 9-3. North American Manufacturers Who Held 5L Licenses

| Manufacturers of API Line Pipe 5L | 3rd Ed Jan-1930 | 4th Ed Jul-1931 | 5th Ed Jan-1934 | 6th Ed Aug-1935 | 7th Ed Apr-1940 | 8th Ed May-1942 | 9th Ed Aug-1944 | 10th Ed Aug-1945 | 11th Ed May-1949 | 12th Ed Mar-1951 | 13th Ed Mar-1954 | 14th Ed Mar-1955 | 15th Ed Mar-1956 | 16th Ed Apr-1957 | 17th Ed Mar-1958 | 18th Ed Feb-1960 | 19th Ed Mar-1962 |
|---|---|---|---|---|---|---|---|---|---|---|---|---|---|---|---|---|---|
| Mobile Pipe Constructors, Inc., Pleasant Hill, CA | | | | | | | | | | | | | | | | | |
| NAPSCO, Bensalem, PA | | | | | | | | | | | | | | | | | |
| National Annealing Box Co., Washington, PA | | | | | | | | | | | | | | | | | |
| National Pipe and Tube Co., Liberty, TX | | | | | | | | | | | | | | | | | |
| National Supply Company, Ambridge, PA (Spang Chalfant Div)[Subsidiary of Armco Steel Corp--17th] | | | | | x | x | x | x | x | x | x | x | x | x | x | x | |
| National Tube Company, Pittsubrgh, PA | x | x | x | | x | x | x | x | x | x | x | x | x | x | x | x | x |
| National Tube Company, Pittsubrgh, PA(United States Steel) | | | | | | | | | | | | | | | | | |
| Newport Steel Corp., Newport,KY | | | | | | | | | | | x | x | x | | | | |
| Northwest Pipe & Casing Co., Clackamas, OR | | | | | | | | | | | | | | | | | |
| Page-Hersey Tubes, Ltd., Ontario, Canada | | | | | x | x | x | x | x | x | x | x | x | x | x | x | x |
| Page-Hersey Tubes Western Ltd. Ontario, Canada | | | | | | | | | | | | | | | | | |
| Paragon Industries, Inc., Sapulpa, OK | | | | | | | | | | | | | | | | | |
| Phoenix Steel Co., Phoenixville, PA | | | | | | | | | | | | | | x | x | x | x |
| Pittsburgh Steel Company, PA | | | | | x | x | x | x | x | x | | | | | | | |
| Pittsburgh Steel Company, Allenport, PA | | | | | | | | | | | | x | x | x | x | x | x |
| Pittsburgh Steel Company, Pittsburgh, PA | | | | | | | | | | | x | | | | | | |
| Pittsburgh Steel Products Company, Pittsburgh, PA | x | x | x | x | | | | | | | | | | | | | |
| Pittsburgh Tube-Darlington/Butler, Monaca, PA | | | | | | | | | | | | | | | | | |
| Prairie Pipe Mfg. Co. Ltd., Regina, Sask., Canada | | | | | | | | | | | | | | | x | x | |
| Productora Mexicana de Tuberia, Mexico, DF [Lazaro Cardenas Michoacan, Mexico] | | | | | | | | | | | | | | | | | |
| Productos Tubulares Monclova, S.A., Monclova, Coah., Mexico | | | | | | | | | | | | | | | | | |
| Prudential Steel Ltd., Calgary, Albeta, Canada | | | | | | | | | | | | | | | | | |
| Quanex Corp., Rosenberg, TX [Belleview, TX / Houston, TX] | | | | | x | | | | | | | | | | | | |
| Ram Steel Corp., Ltd., Red Deer, Albeta, Canada | | | | | | | | | | | | | | | | | |
| Reading Iron Company, Reading, PA | x | x | | x | | | | | | | | | | | | | |

## Table 9-3. North American Manufacturers Who Held 5L Licenses

| Manufacturers of API Line Pipe 5L | 20th Ed Mar-1963 | 21st Ed Mar-1965 | 22nd Ed Mar-1967 | 23rd Ed Mar-1968 | 24th Ed Apr-1969 | 25th Ed Apr-1970 | 26th Ed Apr-1971 | 27th Ed Mar-1973 | 28th Ed Mar-1975 | 29th Ed Mar-1977 | 30th Ed Mar-1978 | 31th Ed Mar-1980 | 32nd Ed Mar-1982 | 33rd Ed Mar-1983 | 34th Ed Mar-1984 | 35th Ed May-1985 | 36th Ed Jun-1987 |
|---|---|---|---|---|---|---|---|---|---|---|---|---|---|---|---|---|---|
| Mobile Pipe Constructors, Inc., Pleasant Hill, CA | | | | | | | | | | | | | | | | | |
| NAPSCO, Bensalem, PA | | | | | | x | x | x | | | | | | x | x | x | x |
| National Annealing Box Co., Washington, PA | | | | | | | | | | | | | | x | x | | |
| National Pipe and Tube Co., Liberty, TX | | | | | | | | | | x | x | x | x | x | x | x | x |
| National Supply Company, Ambridge, PA (Spang Chalfant Div)[Subsidiary of Armco Steel Corp-17th] | | | | | | | | | | | | | | | | | |
| National Tube Company, Pittsburgh, PA | | | | | | | | | | | | | | | | | |
| National Tube Company, Pittsburgh, PA(United States Steel) | x | | | | | | | | | | | | | | | | |
| Newport Steel Corp., Newport, KY | | | | x | x | x | x | x | x | x | x | x | x | x | x | x | x |
| Northwest Pipe & Casing Co., Clackamas, OR | | x | | x | x | x | x | x | x | x | x | x | x | x | x | x | |
| Page-Hersey Tubes, Ltd., Ontario, Canada | x | x | | | | | | | | | | | | | | | |
| Page-Hersey Tubes Western Ltd., Ontario, Canada | x | | | | | | | | | | | | | | | | |
| Paragon Industries, Inc. Sapulpa, OK | | | | | | | | | | | | | | x | x | x | x |
| Phoenix Steel Co., Phoenixville, PA | x | x | x | x | x | x | x | x | x | x | x | x | x | x | x | x | x |
| Pittsburgh Steel Company, Allenport, PA | x | x | x | x | | | | | | | | | | | | | |
| Pittsburgh Steel Company, Pittsburgh, PA | | | | | | | | | | | | | | | | | |
| Pittsburgh Steel Products Company, Pittsburgh, PA | | | | | | | | | | | | | | | | | |
| Pittsburgh Tube-Darlington/Butler, Monaca, PA | | | | | | | | | | | | | | | | x | x |
| Prairie Pipe Mfg. Co. Ltd., Regina, Sask., Canada | | | | | | | | | | | | | | | | | |
| Productora Mexicana de Tuberia, Mexico, DF (Lazaro Cardenas Michoacan, Mexico) | | | | | | | | | | | | | | | | | x |
| Productos Tubulares Monclova, S.A., Monclova, Coah., Mexico | | | | | x | x | x | x | x | x | x | x | x | x | x | x | x |
| Prudential Steel Ltd., Calgary, Alberta, Canada | | | | x | x | x | x | x | x | x | x | x | x | x | x | x | x |
| Quanex Corp., Rosenberg, TX (Belleview, TX / Houston, TX) | | | | | | | | | | | | x | x | x | x | x | x |
| Ram Steel Corp., Ltd., Red Deer, Alberta, Canada | | | | | | | | | | | | | | x | x | | |
| Reading Iron Company, Reading, PA | | | | | | | | | | | | | | | | | |

## Table 9-3. North American Manufacturers Who Held 5L Licenses

| Manufacturers of API Line Pipe 5L | 3rd Ed Jan-1930 | 4th Ed Jul-1931 | 5th Ed Jan-1934 | 6th Ed Aug-1935 | 7th Ed Apr-1940 | 8th Ed May-1942 | 9th Ed Aug-1944 | 10th Ed Aug-1945 | 11th Ed May-1949 | 12th Ed Mar-1951 | 13th Ed Mar-1954 | 14th Ed Mar-1955 | 15th Ed Mar-1956 | 16th Ed Apr-1957 | 17th Ed Mar-1958 | 18th Ed Feb-1960 | 19th Ed Mar-1962 |
|---|---|---|---|---|---|---|---|---|---|---|---|---|---|---|---|---|---|
| Republic Steel Corp., Youngstown, OH | x | x | x | x |  | x | x | x | x | x | x | x | x | x | x | x | x |
| Steel and Tubes Div., Cleveland, OH |  |  |  |  |  |  |  |  |  |  |  |  |  | x | x | x | x |
| R.O. Industries, Newton Falls, OH |  |  |  |  |  |  |  |  |  |  |  |  |  |  |  |  |  |
| Sawhill Tubular Products, Inc., Sharon, PA |  |  |  |  |  |  |  |  |  |  |  |  |  |  |  | x | x |
| Div. of Cyclops Corp. after 24th Ed. |  |  |  |  |  |  |  |  |  |  |  |  |  |  |  |  |  |
| Sidbec-Dosco Ltd., Montreal, Quebec, Canada |  |  |  |  |  |  |  |  |  |  |  |  |  |  |  |  |  |
| Smith, A.O., Corp., Milwaukee, WI |  |  |  |  |  |  |  |  |  |  |  | x | x | x | x | x | x |
| Smith, A.O., Corp. of Texas, Milwaukee, WI |  |  |  |  |  |  |  |  |  |  |  |  |  | x | x | x | x |
| Smith-Scott Co., Inc., Riverside, CA (P-E only) |  |  |  |  |  |  |  |  |  |  |  |  |  |  |  | x | x |
| Sonco Steel Tube Ltd., Brampton, Ontario, Canada |  |  |  |  |  |  |  |  |  |  |  |  |  |  |  |  |  |
| South Chester Tube Co., Chester, PA | x | x | x | x | x | x | x | x | x | x | x | x | x | x |  |  |  |
| Southern Pipe & Casing Co., Azusa, CA |  |  |  |  |  |  |  |  |  |  |  |  |  |  | x |  |  |
| Southern Pipe, Div. of U.S. Industries, Inc., Azusa, CA |  |  |  |  |  |  |  |  |  |  |  |  |  |  |  | x | x |
| Southwestern Pipe. Inc., Houston, TX |  |  |  |  |  |  |  |  |  |  |  |  |  |  |  | x | x |
| Southwestern Pipe of Colorado, Inc., Ft. Collins, CO |  |  |  |  |  |  |  |  |  |  |  |  |  |  |  |  |  |
| Southwest Fabricating & Welding Co., Inc., Houston, TX |  |  |  |  |  |  |  |  |  |  |  |  |  |  |  |  |  |
| Spang, Chalfant & Company, Pittsburgh, PA | x | x | x | x |  |  |  |  |  |  |  |  |  |  |  |  |  |
| Standard Tube and T.I. Limited, Woodstock, Ont., Canada |  |  |  |  |  |  |  |  |  |  |  |  |  |  |  |  | x |
| Standard Tube of Canada, Ltd., Woodstock, Ontario, Canada |  |  |  |  |  |  |  |  |  |  |  |  |  |  |  |  |  |
| Standard Tube Company, Detroit, MI |  |  |  |  |  |  |  |  |  |  |  |  |  |  |  | x | x |
| STELCO Inc., Hamilton, Ontario, Canada |  |  |  |  |  |  |  |  |  |  |  |  |  | x | x | x | x |
| Stelco Pipe & Tube, Welland Ontario, Canada |  |  |  |  |  |  |  |  |  |  |  |  |  |  |  |  |  |
| Camrose Alberta, Canada |  |  |  |  |  |  |  |  |  |  |  |  |  |  |  |  |  |
| Page Hersey Facility, Welland, Ontario, Canada |  |  |  |  |  |  |  |  |  |  |  |  |  |  |  |  |  |

## Table 9-3. North American Manufacturers Who Held 5L Licenses

| Manufacturers of API Line Pipe 5L | 20th Ed Mar-1963 | 21st Ed Mar-1965 | 22nd Ed Mar-1967 | 23rd Ed Mar-1968 | 24th Ed Apr-1969 | 25th Ed Apr-1970 | 26th Ed Apr-1971 | 27th Ed Mar-1973 | 28th Ed Mar-1975 | 29th Ed Mar-1977 | 30th Ed Mar-1978 | 31st Ed Mar-1980 | 32nd Ed Mar-1982 | 33rd Ed Mar-1983 | 34th Ed Mar-1984 | 35th Ed May-1985 | 36th Ed Jun-1987 |
|---|---|---|---|---|---|---|---|---|---|---|---|---|---|---|---|---|---|
| Repbulic Steel Corp., Youngstown, OH (Steel and Tubes Div., Cleveland, OH) | x | x | x | x | x | x | x | x | x |  | x | x | x | x | x |  |  |
| R.O. Industries, Newton Falls, OH |  |  |  |  |  |  |  |  |  | x | x |  |  |  |  |  |  |
| Sawhill Tubular Products, Inc., Sharon, PA (Div of Cyclops Corp. after 24th Ed.) |  | x | x | x |  |  |  |  |  |  |  |  |  |  | x | x |  |
| Sidbec-Dosco Ltd., Montreal, Quebec, Canada |  |  |  |  | x | x | x | x | x | x | x | x | x | x | x | x | x |
| Smith, A.O., Corp., Milwaukee, WI |  | x | x | x | x | x | x | x | x | x | x | x | x | x | x | x |  |
| Smith, A.O., Corp. of Texas, Milwaukee, WI | x | x | x | x | x | x | x | x |  |  |  |  |  |  |  |  |  |
| Smith-Scott Co., Inc., Riverside, CA (P-E only) | x | x | x | x |  |  |  |  |  |  |  |  |  |  |  |  |  |
| Sonco Steel Tube Ltd., Brampton, Ontario, Canada |  |  |  |  |  |  |  |  |  |  |  |  |  | x | x | x | x |
| South Chester Tube Co., Chester, PA |  |  |  |  |  |  |  |  | x | x | x | x | x | x |  |  |  |
| Southern Pipe & Casing Co., Azusa, CA |  | x | x | x | x |  |  |  |  |  |  |  |  |  |  |  |  |
| Southern Pipe, Div., of U.S. Industries, Inc., Azusa, CA |  |  |  |  | x |  |  |  |  |  |  |  |  |  |  |  |  |
| Southwestern Pipe, Inc., Houston, TX | x | x | x | x | x | x | x | x |  |  |  |  |  |  |  |  |  |
| Southwestern Pipe of Colorado, Inc. Ft. Collins, CO |  | x | x | x | x | x |  |  |  |  |  |  |  |  |  |  |  |
| Southwest Fabricating & Welding Co., Inc., Houston, TX |  |  |  |  |  |  |  |  |  |  |  |  |  | x | x | x |  |
| Spang, Chalfant & Company, Pittsburgh, PA |  |  |  |  |  |  |  |  |  |  |  |  |  |  |  |  |  |
| Standard Tube and T.I. Limited, Woodstock, Ont., Canada | x | x | x | x | x |  |  |  |  |  |  |  |  |  |  |  |  |
| Standard Tube of Canada, Ltd., Woodstock, Ontario, Canada |  |  |  |  |  | x | x | x | x |  |  |  |  |  |  |  |  |
| Standard Tube Company, Detroit, MI | x | x | x | x | x | x | x | x | x | x | x | x | x | x | x | x |  |
| STELCO Inc., Hamilton, Ontario, Canada | x | x | x | x | x | x | x | x | x | x | x | x | x | x | x | x |  |
| Stelco Pipe & Tube, Welland Ontario, Canada |  |  |  |  |  |  |  |  |  |  |  |  |  |  |  |  | x |
| Camrose Alberta, Canada |  |  |  |  |  |  |  |  |  |  |  |  |  |  |  |  | x |
| Page Hersey Facility, Welland, Ontario, Canada |  |  |  |  |  |  |  |  |  |  |  |  |  |  |  |  | x |

Table 9-3. North American Manufacturers Who Held 5L Licenses

| Manufacturers of API Line Pipe — 5L | 3rd Ed Jan-1930 | 4th Ed Jul-1931 | 5th Ed Jan-1934 | 6th Ed Aug-1935 | 7th Ed Apr-1940 | 8th Ed May-1942 | 9th Ed Aug-1944 | 10th Ed Aug-1945 | 11th Ed May-1949 | 12th Ed Mar-1951 | 13th Ed Mar-1954 | 14th Ed Mar-1955 | 15th Ed Mar-1956 | 16th Ed Apr-1957 | 17th Ed Mar-1958 | 18th Ed Feb-1960 | 19th Ed Mar-1962 |
|---|---|---|---|---|---|---|---|---|---|---|---|---|---|---|---|---|---|
| Stupp Corporation, Baton Rouge, LA | | | | | | | | | | | | | | | | | |
| Taylor Forge 7 Pipe Works, Chicago, IL | | | | | x | x | x | x | x | x | x | x | | | | | |
| Teledyne Pipe, Galveston, TX | | | | | | | | | | | | | | | | | |
| Texas Pipe Bending Co., Houston, TX | | | | | | | | | | | | | | | | | |
| Tex-Tube Inc., Houston, Texas | | | | | | | | | | | | | | | x | x | x |
| Thompson Pipe & Steel Co., Denver, CO | | | | | | | | | | | | | | | | | |
| Timberline Tube Inc., Fort Collins, CO (Southwestern Pipe of Colorado, Inc.) | | | | | | | | | | | | | | | | | |
| Trans American Pipe Corp., Bensalem, PA | | | | | | | | | | | | | | | | | |
| TUBESA, S.A., Mexico | | | | | | | | | | | | | | | | | |
| Tubacero, S.A., Monterrey, N.L., Mexico | | | | | | | | | | | | | | | x | x | x |
| Tuberia Laguna, S.A., Gomez Palacio, DGO, Mexico | | | | | | | | | | | | | | | | | |
| Tuberia y Estructuras, S.A., Mexico D.F. | | | | | | | | | | | | | | | | | |
| Tubos de Acero de Mexico, S.A., Mexico, D.F. | | | | | | | | | | | | | x | x | x | x | x |
| Tubular Finishing Works, Inc., Navasta, TX | | | | | | | | | | | | | | | | | |
| United Concrete Pipe Corp., Baldwin Park, CA | | | | | | | | | | | | | | | | | |
| United States Steel Corp., Pittsburgh, PA | | | | | | | | | | | | | | | | | |
| Fairfield, AL | | | | | | | | | | | | | | | | | |
| Geneva Facility Provo | | | | | | | | | | | | | | | | | |
| Lorain, OH | | | | | | | | | | | | | | | | | |
| Baytown, TX | | | | | | | | | | | | | | | | | |
| Gary, IN | | | | | | | | | | | | | | | | | |
| Fairless Hill, PA | | | | | | | | | | | | | | | | | |
| U.S. Industries, Inc. Tubular Prod. Div., Azusa, CA | | | | | | | | | | | | | | | | | |
| USS, Pittsburgh, PA | | | | | | | | | | | | | | | | | |
| Valley Manufacturing Co., Valley, NE | | | | | | | | | | | | | | | | | |
| Valmont Industries, Inc., Valley, NE | | | | | | | | | | | | | | | | | x |
| Western Pipe & Steel Co. of Calif., San Francisco, CA | | | | | x | | x | x | | | | | | | | | |
| Wheatland Tube Co., Philadelphia, PA | | | | | | | | | | | | | x | x | x | x | x |
| Wheeling-Pittsburgh Steel Corp., Allenport, PA | | | | | | | | | | | | | | | | | |
| Monessen, PA | | | | | | | | | | | | | | | | | |
| Wheeling, WV | | | | | | | | | | | | | | | | | |

## Table 9-3. North American Manufacturers Who Held 5L Licenses

| Manufacturers of API Line Pipe 5L | 20th Ed Mar-1963 | 21st Ed Mar-1965 | 22nd Ed Mar-1967 | 23rd Ed Mar-1968 | 24th Ed Apr-1969 | 25th Ed Apr-1970 | 26th Ed Apr-1971 | 27th Ed Mar-1973 | 28th Ed Mar-1975 | 29th Ed Mar-1977 | 30th Ed Mar-1978 | 31th Ed Mar-1980 | 32nd Ed Mar-1982 | 33rd Ed Mar-1983 | 34th Ed Mar-1984 | 35th Ed May-1985 | 36th Ed Jun-1987 |
|---|---|---|---|---|---|---|---|---|---|---|---|---|---|---|---|---|---|
| Stupp Corporation, Baton Rouge, LA | x | x | x | x | x | x | x | x | x | x | x | x | x | x | x | x | x |
| Taylor Forge 7 Pipe Works, Chicago, IL | | | | | | | | | | | | | | | | | |
| Teledyne Pipe, Galveston, TX | | | | | | | | | x | x | x | x | x | x | x | x | |
| Texas Pipe Bending Co., Houston, TX | | | | | | | | | x | | x | x | x | x | x | x | |
| Tex-Tube Inc., Houston, Texas | x | x | x | x | x | x | | x | x | x | | | | | | | |
| Thompson Pipe & Steel Co., Denver, CO | | | | | | | | | | | | | | | | | x |
| Timberline Tube Inc., Fort Collins, CO (Southwestern Pipe of Colorado, Inc.) | x | | | | | | | | | | | | | | | | |
| Trans American Pipe Corp., Bensalem, PA | | | | | | | | | | | | | | | | | |
| TUBESA, S.A., Mexico | x | x | x | x | x | x | x | x | x | x | x | x | x | x | x | x | x |
| Tubacero, S.A., Monterrey, N.L., Mexico | x | x | x | x | x | x | x | x | x | x | x | x | x | x | x | x | x |
| Tuberia Laguna, S.A., Gomez Palacio, DGO, Mexico | | | | | | | | | | | x | x | x | x | x | x | |
| Tuberia y Estructuras, S.A., Mexico D.F. | | | | | | | | | | | | | | x | x | | |
| Tubos de Acero de Mexico, S.A., Mexico, D.F. | x | x | x | x | x | x | | x | x | x | x | x | x | | x | x | x |
| Tubular Finishing Works, Inc., Navasta, TX | | | | | | | | | | | x | x(EF) | | | | | |
| United Concrete Pipe Corp., Baldwin Park, CA | | | | | | x | | | | | | | | | | | |
| United States Steel Corp., Pittsburgh, PA | x | x | x | x | x | x | x | x | x | x | x | x | x | x | x | x | x |
| United States Steel Corp. — Fairfield, AL | | | | | | | | | | | | | | | | | x |
| United States Steel Corp. — Geneva Facility Provo | | | | | | | | | | | | | | | | x | x |
| United States Steel Corp. — Lorain, OH | | | | | | | | | | | | | | | | x | x |
| United States Steel Corp. — Baytown, TX | | | | | | | | | | | | | | | x | x | x |
| United States Steel Corp. — Gary, IN | | | | | | | | | | | | | | | | x | x |
| United States Steel Corp. — Fairless Hill, PA | | | | | | | | | | | | | | | | x | x |
| U S Industries, Inc. Tubular Prod Div., Azusa, CA | x | | | | | | | | | | | | | | | | x |
| USS, Pittsburgh, PA | | | | | | | | | | | | | | | | | |
| Valley Manufacturing Co., Valley, NE | x | x | x | | | | | | | | | | | | | | |
| Valmont Industries, Inc., Valley, NE | | | | x | | | | x | x | x | x | x | x | x | x | x | x |
| Western Pipe & Steel Co. of Calif., San Francisco, CA | x | x | x | x | x | x | x | x | x | x | x | | | | | | |
| Wheatland Tube Co., Philadelphia, PA | x | x | x | x | x | x | x | x | x | x | x | x | x | x | x | x | |
| Wheeling-Pittsburgh Steel Corp., Allenport, PA | | | | | | x | | | | | | | | | x | x | |
| Wheeling-Pittsburgh Steel Corp. — Monessen, PA | | | | | x | | | | | | x | | x | | | | |
| Wheeling-Pittsburgh Steel Corp. — Wheeling, WV | | | | | | | | | | | | x | x | | | | |

## Table 9-3. North American Manufacturers Who Held 5L Licenses

| Manufacturers of API Line Pipe 5L | 3rd Ed Jan-1930 | 4th Ed Jul-1931 | 5th Ed Jan-1934 | 6th Ed Aug-1935 | 7th Ed Apr-1940 | 8th Ed May-1942 | 9th Ed Aug-1944 | 10th Ed Aug-1945 | 11th Ed May-1949 | 12th Ed Mar-1951 | 13th Ed Mar-1954 | 14th Ed Mar-1955 | 15th Ed Mar-1956 | 16th Ed Apr-1957 | 17th Ed Mar-1958 | 18th Ed Feb-1960 | 19th Ed Mar-1962 |
|---|---|---|---|---|---|---|---|---|---|---|---|---|---|---|---|---|---|
| Wheeling Steel Corporation, Benwood, WV | | | | | | | | | | | | x | x | x | x | x | x |
| Wheeling Steel Corporation, Wheeling, WV | x | x | x | x | x | x | x | x | x | x | x | | | | | | |
| Youngstown Sheet & Tube Company, Youngstown, OH | x | x | x | x | x | x | x | x | x | x | x | x | x | x | x | x | x |

**Table 9-3. North American Manufacturers Who Held 5L Licenses**

| Manufacturers of API Line Pipe | 20th Ed Mar-1963 | 21st Ed Mar-1965 | 22nd Ed Mar-1967 | 23rd Ed Mar-1968 | 24th Ed Apr-1969 | 25th Ed Apr-1970 | 26th Ed Apr-1971 | 27th Ed Mar-1973 | 28th Ed Mar-1975 | 29th Ed Mar-1977 | 30th Ed Mar-1978 | 31th Ed Mar-1980 | 32nd Ed Mar-1982 | 33rd Ed Mar-1983 | 34th Ed Mar-1984 | 35th Ed May-1985 | 36th Ed Jun-1987 |
|---|---|---|---|---|---|---|---|---|---|---|---|---|---|---|---|---|---|
| 5L | | | | | | | | | | | | | | | | | |
| Wheeling Steel Corporation, Benwood, WV | x | | x | x | | | | | | | | | | | | | |
| Wheeling Steel Corporation, Wheeling, WV | | | | | | | | | | | | | | | | | |
| Youngstown Sheet & Tube Company, Youngstown, OH | x | x | x | x | x | x | x | x | x | | x | | | | | | |

Table 9-4. North American Manufacturers Who Held 5LX Licenses

| Manufacturers of API Line Pipe / 5LX | 3rd Ed Mar-1951 | 4th Ed Mar-1953 | 5th Ed Nov-1954 | 6th Ed Feb-1956 | 7th Ed Apr-1957 | 8th Ed Mar-1958 | 9th Ed Feb-1960 | 10th Ed Mar-1962 | 11th Ed Mar-1963 | 12th Ed Mar-1965 | 13th Ed Mar-1966 | 14th Ed Mar-1967 | 15th Ed Mar-1968 | 16th Ed Apr-1969 | 17th Ed Apr-1970 | 18th Ed Apr-1971 | 19th Ed Mar-1973 | 20th Ed Mar-1975 | 21st Ed Mar-1977 | 22nd Ed Mar-1978 | 23rd Ed Mar-1980 | 24th Ed Mar-1982 |
|---|---|---|---|---|---|---|---|---|---|---|---|---|---|---|---|---|---|---|---|---|---|---|
| Acme-Newport Steel Co., New Port, KY | | | | | x | x | x | x | x | | | | | | | | | | | | | |
| Alberta Phoenix Tube & Pipe Ltd., Edmonton, Alberta, Canada | | | | | x | x | x | x | x | x | x | x | | | | | | | | | | |
| Algoma Steel Corp., Ltd., Sault Ste. Marie, Canada | | | | | | | | | | | | | | | | x | x | x | x | x | x | x |
| American Bridge Div, Pittsburgh, PA (United States Steel) | | | | x | x | x | x | x | x | x | x | x | x | x | x | x | x | x | x | x | x | x |
| American Cast Iron Pipe Co., Birmingham, AL — American Steel Pipe Div | | | | | | | | | | | x | | | | | | | | | | | |
| American Pipe & Construction Co., Portland, OR | | | | | | | | | x | x | x | x | x | x | x | x | | | | | | |
| Ameron, Portland, OR | | | | | | | | | | | | | | | x | | | | | | | |
| Steel Fabricating Division | | | | | | | | | | | | | | | | x | | | | | | |
| Armco Steel Corp., Middletown, OH — Seamless Tubular, Houston, TX | | | | | | | | | | | | | | | | x | x | x | x | x | x | x |
| Armco Div., Ambridge, PA | | | | | | | | | | | | | | | | | x | | | | | |
| Beall Pipe and Tank Corp., Portland, OR | | | | x | x | x | x | x | x | x | x | x | x | x | x | x | x | x | x | x | x | x |
| Berg Steel Pipe Corp., Panama City, FL | | | | | | | | | | | | | | | | | | | | | x | x |
| Bethlehem Steel Co., Bethlehem, PA | x | x | x | x | x | x | x | x | x | x | x | x | x | x | x | x | x | x | x | x | x | x |
| Big Inch Pipe Corp Ltd., Calgary, Alta. Canada | | | | | | | x | x | x | | | | | | | | | | | | | |
| Brooks Tube Ltd., Brooks, Alberta, Canada | | | | | | | | | | | | | | | | | | x | x | x | x | x |
| Bull Moose Tube Co., Gerald, MO | | | | | | | | | | | | | | | | | | | | | x | x |
| Cal-Metal Pipe Corp of Louisiana, Baton Rouge, LA | | | | x | | x | x | | | | | | | | | | | | | | | |

9- 85

Table 9-4. North American Manufacturers Who Held 5LX Licenses

| Manufacturers of API Line Pipe / 5LX | 3rd Ed Mar-1951 | 4th Ed Mar-1953 | 5th Ed Nov-1954 | 6th Ed Feb-1956 | 7th Ed Apr-1957 | 8th Ed Mar-1958 | 9th Ed Feb-1960 | 10th Ed Mar-1962 | 11th Ed Mar-1963 | 12th Ed Mar-1965 | 13th Ed Mar-1966 | 14th Ed Mar-1967 | 15th Ed Mar-1968 | 16th Ed Apr-1969 | 17th Ed Apr-1970 | 18th Ed Apr-1971 | 19th Ed Mar-1973 | 20th Ed Mar-1975 | 21st Ed Mar-1977 | 22nd Ed Mar-1978 | 23rd Ed Mar-1980 | 24th Ed Mar-1982 |
|---|---|---|---|---|---|---|---|---|---|---|---|---|---|---|---|---|---|---|---|---|---|---|
| Cal-Metal Corp., Torrance, CA | | | | | | | x | x | x | x | x | x | x | x | x | x | x | x | x | x | | x |
| Cameron Iron Works, Inc., Houston, TX | | | | | | | | | | | | | x | x | x | x | x | x | x | x | x | x |
| Camrose Tubes Ltd., Toronto, Ontario, Canada | | | | | | | x | x | x | x | | | | | | | | | | | | |
| Canadian Phoenix Steel & Pipe, Ltd., Edmonton, Alberta, Canada | | | | | | | | | | | | | x | x | x | x | x | | | | | |
| Canadian Western Pipe Mills, Ltd., Port Moody, B.C., Canada | | | | | | | | | x | x | x | x | | | | | | | | | | |
| Chemetron Corp., Louisville, KY (Tube Turns Div) | | | | | | | | | | | | | | | | x | x | x | x | x | x | x |
| Cherokee Steel, Inc., Tulsa, OK | | | | | | | | x | x | x | | | | | | | | | | | | |
| Claymont Steel Corp., Claymont, DE | x | | | | | | | | | | | | | | | | | | | | | |
| Colorado Fuel & Iron Corp., Pueblo, CO | | | | | x | x | x | x | x | x | x | x | x | x | x | x | x | x | x | x | x | x |
| Colorado Fuel & Iron Corp., Pueblo, CO (Wickwire Spencer Steel Div., Claymont, DE) | | x | x | x | | | | | | | | | | | | | | | | | | |
| Consolidated Western Steel Corp. Los Angeles, CA (United States Steel Corp.) — San Francisco, CA | x | x | x | x | x | x | x | x | x | | | | | | | | | | | | | |
| (Houston, TX) | x | x | x | x | x | x | x | x | x | | | | | | | | | | | | | |
| (Berkely, CA) | | x | x | x | x | x | x | x | x | | | | | | | | | | | | | |
| Cyclops Corp., Houston, TX (Tex-Tube Div) | | | | | | | | | | | | | | | | | | | | x | x | x |
| Donovan Steel Tube Co., Toledo, OH | | | | | | | | | | | | | | | | | | | | | | x |
| Fort Worth Pipe & Supply, Fort Worth, TX | | | | | | | | | | | | | | | | | | x | x | | x | x |
| Fox Steel Pipe Corp., Jacksonville, FL | | | | | | | x | x | | | | | | | | | | | | | | |
| Gulf-Western Manufacturing Co., Chicago, IL (G-W Energy Products Group) | | | | | | | | | | | | | | | | | | | | x | | |

Table 9-4. North American Manufacturers Who Held 5LX Licenses

| Manufacturers of API Line Pipe 5LX | 3rd Ed Mar-1951 | 4th Ed Mar-1953 | 5th Ed Nov-1954 | 6th Ed Feb-1956 | 7th Ed Apr-1957 | 8th Ed Mar-1958 | 9th Ed Feb-1960 | 10th Ed Mar-1962 | 11th Ed Mar-1963 | 12th Ed Mar-1965 | 13th Ed Mar-1966 | 14th Ed Mar-1967 | 15th Ed Mar-1968 | 16th Ed Apr-1969 | 17th Ed Apr-1970 | 18th Ed Apr-1971 | 19th Ed Mar-1973 | 20th Ed Mar-1975 | 21st Ed Mar-1977 | 22nd Ed Mar-1978 | 23rd Ed Mar-1980 | 24th Ed Mar-1982 |
|---|---|---|---|---|---|---|---|---|---|---|---|---|---|---|---|---|---|---|---|---|---|---|
| HYLSA, S.A. Monterrey, N.L. Mexico | | | | | | | | | | | | | | | | | | | X | X | X | X |
| Interlake Steel Corp., Newport, KY | | | | | | | | | | | | | | | | | | | | X | | |
| International Portable Pipe Mills Ltd., Calgary, Alberta, Canada | | | | | | | | | | | | | | | | | | X | | | | |
| Interprovincial Steel and Pipe Corp. Ltd. Regina, Sask. Canada | | | | | | | X | X | X | X | X | X | X | X | X | X | X | X | X | X | X | X |
| Jones & Laughlin Steel Corporation, Aliquippa, PA | X | X | X | X | X | X | X | X | X | X | X | X | X | X | X | X | X | X | X | X | X | X |
| Kaiser Steel Corp. Fontana, CA | X | X | X | X | X | X | X | X | X | X | X | X | X | X | X | X | X | X | X | X | X | X |
| Oakland, CA | | | | | | | | | | | | | | | | | | | | | | |
| Kane Boiler Works, Inc. Galveston, TX | | | | | | | X | X | X | X | X | X | X | X | X | X | X | X | | | | |
| Kane Industries, Inc. Galveston, TX | | | | | | | | | | | | | | | | | | | | | X | X |
| Lone Star Steel Co. Dallas, TX | | X | X | X | X | X | X | X | X | X | X | X | X | X | X | X | | X | | X | X | X |
| Mannesmann Tube Co. Ltd., Sault Ste. Marie, Ont. Canada | | | | | | X | X | X | X | X | | X | | X | X | : | | | | | | |
| Mario Maraldi S.p.A. Forlimpopoli, Forli, Italy | | | | | | X | | | | | | | X | | X | | | | | | | |
| Maruichi American Corp. Santa Fe Springs, CA | | | | | | | | | | | | | | | | | | | | | X | X |
| Master Tank & Welding, Dallas, TX | X | X | X | X | X | X | X | X | X | X | | | | | | | | | | | | |
| Maverick Tube Corp., St. Louis, MO | | | | | | | | | | | | | | | | | | | | | X | X |
| McNamar Boiler & Tank Co. Tulsa, OK | X | X | X | X | X | | | | | | | | | | | | | | | | | |
| Mobile Pipe Constructors, Inc. Pleasant Hill, CA | | | | | | | | | | | | | | | X | X | X | | | | | |
| National Pipe & Tube Co. Liberty, TX | | | | | | | | | | | | | | | | | | | X | X | X | X |

## Table 9-4. North American Manufacturers Who Held 5LX Licenses

| Manufacturers of API Line Pipe 5LX | 3rd Ed Mar-1951 | 4th Ed Mar-1953 | 5th Ed Nov-1954 | 6th Ed Feb-1956 | 7th Ed Apr-1957 | 8th Ed Mar-1958 | 9th Ed Feb-1960 | 10th Ed Mar-1962 | 11th Ed Mar-1963 | 12th Ed Mar-1965 | 13th Ed Mar-1966 | 14th Ed Mar-1967 | 15th Ed Mar-1968 | 16th Ed Apr-1969 | 17th Ed Apr-1970 | 18th Ed Apr-1971 | 19th Ed Mar-1973 | 20th Ed Mar-1975 | 21st Ed Mar-1977 | 22nd Ed Mar-1978 | 23rd Ed Mar-1980 | 24th Ed Mar-1982 |
|---|---|---|---|---|---|---|---|---|---|---|---|---|---|---|---|---|---|---|---|---|---|---|
| National Supply Company, Ambridge, PA (Spang Chalfant Div)(Subsidiary of Armco Steel Corp) | | | | | | | | | | | | | | | | | | | | | | |
| National Tube Company, Pittsburgh, PA | x | x | x | x | x | x | x | x | x | | | | | | | | | | | | | |
| Newport Steel Corp., Newport KY | | x | x | x | | | | | | | | | | | | | | | | | | x |
| Northwest Pipe & Casing Co., Clackamas, OR | | | | | | | | | | | | | x | x | x | x | x | x | x | x | x | x |
| Page-Hersey Tubes, Ltd., Ontario, Canada | x | | x | x | x | x | x | x | x | | | | | | | | | | | | | |
| Page-Hersey Tubes Western Ltd., Ontario, Canada | | | | | | | | | x | | | | | | | | | | | | | |
| Paragon Pipe Co., Sapulpa, OK | | | | | | | | | | | | | | | | | | | | | | |
| Phoenix Steel Corp., Phoenixville, PA | | | | | x | x | x | x | x | x | x | x | x | x | x | x | x | x | x | x | x | x |
| Prairie Pipe Mfg Co. Ltd, Regina, Sask., Canada | | | | | | x | • | | | | | | | | | | | | | | | |
| Productos Tubulares Monclova, S.A., Monclova, Coah., Mexico | | | | | | | | | | | x | x | x | x | x | x | x | x | x | x | x | x |
| Prudential Steel Ltd., Calgary, Alberta, Canada | | | | | | | | | | | x | x | x | x | x | x | x | x | x | x | x | x |
| Republic Steel Corp., Youngstown, OH [Steel and Tubes Div., Cleveland, OH] | x | x | | | | | | | | | x | x | x | x | x | x | x | x | x | x | | |
| Smith, A.O., Corp., Milwaukee, WI | | | x | x | x | x | x | x | x | x | x | x | x | x | x | x | x | | | | | |
| Smith, A.O., Corp of Texas, Milwaukee, WI | | | x | x | x | x | x | x | x | x | x | x | x | x | x | x | x | | | | | |
| Smith-Scott Co., Inc., Riverside, CA | | | | | | | | | | x | x | x | x | x | | | | | | | | |
| Smith Industries, Inc., Houston, TX | | | | | | | | | | | | | | | | | x | x | x | x | x | x |
| Southern Pipe & Casing Co., Azusa, CA | | | | | | x | x | | | | | | | | | | | | | | | |

## Table 9-4. North American Manufacturers Who Held 5LX Licenses

| Manufacturers of API Line Pipe — 5LX | 3rd Ed Mar-1951 | 4th Ed Mar-1953 | 5th Ed Nov-1954 | 6th Ed Feb-1956 | 7th Ed Apr-1957 | 8th Ed Mar-1958 | 9th Ed Feb-1960 | 10th Ed Mar-1962 | 11th Ed Mar-1963 | 12th Ed Mar-1965 | 13th Ed Mar-1966 | 14th Ed Mar-1967 | 15th Ed Mar-1968 | 16th Ed Apr-1969 | 17th Ed Apr-1970 | 18th Ed Apr-1971 | 19th Ed Mar-1973 | 20th Ed Mar-1975 | 21st Ed Mar-1977 | 22nd Ed Mar-1978 | 23rd Ed Mar-1980 | 24th Ed Mar-1982 |
|---|---|---|---|---|---|---|---|---|---|---|---|---|---|---|---|---|---|---|---|---|---|---|
| (Div of American Pipe & Const Co.) | | | | | | | | | | | | | | | | | | | | | | |
| Southern Pipe, Div of U S Industries, Inc., Azusa, CA | | | | | | | | | | x | x | x | x | x | | | | | | | | |
| Southwestern Pipe, Inc., Houston, TX | | | | | | | | | x | x | x | x | x | x | x | x | x | | | | | |
| Southwestern Pipe of Colorado, Inc., Ft Collins, CO | | | | | | | | | | | | x | x | x | | | | | | | | |
| STELCO Inc., Hamilton, Ontario, Canada | | | | | | | | | x | x | x | x | x | x | x | x | x | x | x | x | x | x |
| Stupp Corporation, Baton Rouge, LA | | | | | | | | x | x | x | x | x | x | x | x | x | x | x | x | x | x | x |
| Taylor Forge Inc., Chicago, IL | | | x | x | x | x | x | x | x | x | x | x | x | x | | | | | | | | |
| Taylor Forge Div (Gulf & Western Industrial Prod., Co., Chicago, IL) | | | | | | | | | | | | | | | x | x | x | | | | | |
| Teledyne Pipe, Galveston, TX | | | | | | | | | | | | | | | | | | x | x | x | x | x |
| Texas Pipe Bending Co., Houston, TX | | | | | | | | | | | | | | | | | | | | | x | x |
| Tex-Tube Inc., Houston, TX (Div of Detroit Steel Corp) | | | | | | x | | x | x | x | x | x | x | x | x | x | x | x | x | | | |
| Timberline Tube Inc., Fort Collins, CO (Southwestern Pipe of Colorado, Inc.) | | | | | | | | | x | | | | | | | | | | | | | |
| Trinity Industries, Inc., Dallas, TX | | | | | | | | | | | x | x | x | x | x | x | x | | | | | |
| Trans American Pipe Corp., Bensalem, PA | | | | | | | | | | | | | | | | | | | | | | x |
| Tubacero, S.A., Monterrey, N.L., Mexico | | | | | | | x | x | x | x | x | x | x | x | x | x | x | | | | | |
| Tuberia Laguna, S.A., Gomez Palacio, DGO, Mexico | | | | | | | | | | | | | | | | | | | | x | x | x |
| Tubos de Acero, S.A., Manufacturera de. Monterrey, NL | | | | | | x | | | | | | | | | | | | | | | | |

## Table 9-4. North American Manufacturers Who Held 5LX Licenses

| Manufacturers of API Line Pipe 5LX | 3rd Ed Mar-1951 | 4th Ed Mar-1953 | 5th Ed Nov-1954 | 6th Ed Feb-1956 | 7th Ed Apr-1957 | 8th Ed Mar-1958 | 9th Ed Feb-1960 | 10th Ed Mar-1962 | 11th Ed Mar-1963 | 12th Ed Mar-1965 | 13th Ed Mar-1966 | 14th Ed Mar-1967 | 15th Ed Mar-1968 | 16th Ed Apr-1969 | 17th Ed Apr-1970 | 18th Ed Apr-1971 | 19th Ed Mar-1973 | 20th Ed Mar-1975 | 21st Ed Mar-1977 | 22nd Ed Mar-1978 | 23rd Ed Mar-1980 | 24th Ed Mar-1982 |
|---|---|---|---|---|---|---|---|---|---|---|---|---|---|---|---|---|---|---|---|---|---|---|
| Tubos de Acero de Mexico, S.A., Mexico, D.F. | | | | | x | x | x | x | x | x | x | x | x | x | x | x | x | x | x | x | x | x |
| United Concrete Pipe Corp., Baldwin Park, CA | | | | | | | | | | | | | | x | x | . | | | | | | |
| United States Steel Corp., Pittsburgh, PA | | | | | | | | | | x | x | x | x | x | x | x | x | x | x | x | x | x |
| U.S. Industries, Inc. Tubular Prod Div, Azusa, CA | | | | | | | | | x | | | | | | | | | | | | | |
| Valley Manufacturing Co., Valley, NE | | | | | | | x | x | x | x | x | | | | | | | | | | | |
| Valmont Industries, Gary, IN Inc., Valley, NE | | | | | | | | | | | | x | x | x | x | x | x | x | x | x | x | x |
| Welland Tubes Limited, Toronto, Ontario, Canada | | | | | x | x | x | x | x | | | | | | | | | | | | | |
| Youngstown Sheet & Tube Company, Youngstown, OH | x | x | x | x | x | x | x | x | x | x | x | x | x | x | x | x | x | x | x | x | | |

**Table 9-5. North American Manufacturers Who Hold Current 5L Licenses (Compiled from "April 1, 1995 Composite List of Manufacturers Licensed for Use of the API Monogram")**

| Manufacturer | Location | CW | Seamless | ERW | DSAW | Spiral DSAW | Unknown |
|---|---|---|---|---|---|---|---|
| Algoma Steel Inc | Sault Ste. Marie Ontario Canada | | X | | | | |
| American Steel Pipe | Birmingham AL | | | X | | | |
| Belville Tube Corp | Bellville TX | | | X | | | |
| Berg Steel Pipe Corp | Panama City FL | | | | X | | |
| California Steel Industries, Inc, Tubular Products | Fontana CA | | | X | | | |
| Camp-Hill Corp | McKeesport PA | | | X | | | |
| Camrose Pipe Co | Camrose Alberta Canada | | | X | X | | |
| CF&I Steel, L.P. | Pueblo CO | | X | | | | |
| Geneva Steel | Vineyard UT | | | X | | | |
| Hylsa SA de CV | San Nicolas de los Garza Mexico | | | X | | | |
| Ingenieria Mecanica Tubular SA de CV | Tialnepantla Mexico | | | | | | X |
| Ipsco, Inc, Calgary Facility | Calgary Alberta Canada | | | X | | | |
| Ipsco, Inc, Edmonton Facility | Edmonton Alberta Canada | | | X | | X | |
| Ipsco, Inc, Red Deer Facility | Red Deer Alberta Canada | | | X | | | |
| Ipsco, Inc., Regina Facility | Regina Saskatchewan Canada | | | X | | X | |
| Ipsco Steel, Inc, Camanche Facility | Camanche IA | | X | | | | |
| Koppel Steel Corp | Ambridge PA | | | X | | | |
| Lone Star Steel Co, Hwy 259 Facility | Lone Star TX | | | X | | | |
| Lone Star Steel Co, Texas Tubular Facility | Lone Star TX | | | X | | | |
| LTV Steel Tubular Products Co, Cleveland Facility | Cleveland OH | | | X | | | |
| LTV Steel Tubular Products Co, Counce Facility | Counce TN | | | X | | | |
| LTV Steel Tubular Products Co, Youngstown Facility | Youngstown OH | | | X | | | |
| Maverick Tube Corp, Conroe Facility | Conroe TX | | | X | | | |
| Maverick Tube Corp, Hickman Facility | Hickman AR | | | X | | | |
| Napa Pipe Corp | Napa CA | | | | X | | |
| Newport Steel Corp | Newport KA | | | X | | | |
| North Star Steel, Houston Facility | Houston TX | | X | | | | |
| North Star Steel, Youngstown Facility | Youngstown OH | | X | | | | |
| Northwest Pipe & Casing of KS | Atchicon KS | | | X | | | |
| Palmer Tube Mills, Inc | Chicago IL | | | X | | | |
| Paragon Industries, Inc | Sapulpa OK | | | | | | X |
| PA Steel Technologies, Inc Steelton Facility | Steelton PA | | | | X | | |
| Pittsburgh Tube Co | Darlington PA | | | X | | | |
| Procarsa, SA de CV | Cuidad Frontera Mexico | | | X | | | |
| Productora Mexicana de Tuberia, SA de CV | Lazaro Cardenas Mexico | | | | X | | |
| Prudential Steel, Ltd | Calgary Alberta Canada | | | X | | | |

Table 9-5. North American Manufacturers Who Hold Current 5L Licenses (Compiled from "April 1, 1995 Composite List of Manufacturers Licensed for Use of the API Monogram")

| Manufacturer | Location | CW | Seamless | ERW | DSAW | Spiral DSAW | Unknown |
|---|---|---|---|---|---|---|---|
| Quanex Corp, Rosenberg Facility | Rosenberg TX | | X | | | | |
| Saw Pipes USA Inc | Baytown TX | | | | X | | |
| Steel Forgings, Inc | Shreveport LA | | | | | | X |
| Stelpipe Ltd, Page-Hersey Facility | Welland Ontario Canada | | | X | X | X | |
| Stupp Corp | Baton Rouge LA | | | X | | | |
| Talleres Acerorey SA | Monterrey Mexico | | | | X | | |
| Tex-Tube Co | Houston TX | | | X | | | |
| Tubacero SA | Monterrey Mexico | | | X | X | | |
| Tuberia Laguna SA de CV | Gomez Palacio Mexico | | | X | | | |
| Tubesa SA de CV | San Luis Potosi Mexico | | | | | X | |
| Tubos de Acero de Mexico SA, Tamsa Facility | Veracruz Mexico | | X | | | | |
| USS, Fairfield Tubular Works | Fairfield AL | | X | | | | |
| USS, Lorain Facility | Lorain OH | | X | | | | |
| USS, McKeesport Facility | McKeesport PA | | | X | | | |
| Villacero Tuberia Nacional SA de CV | Nuevo Leon Mexico | | | X | | | |
| Welland Pipes Ltd, Welland Tube Works | Welland Ontario Canada | | | | X | X | |
| Wheatland Tube Company | Wheatland PA | X | | | | | |

# Glossary of Terms and Definitions

(Taken whole or in part from several references including: Pittsburgh Steel's 1944 Catalog, "The Making, Shaping and Treating of Steel", API Bulletin 5T1, Ninth Edition, May 31, 1988, API Specification 5L, Various Editions, ASM Metals handbook, Various Editions, Various National Tube Company Catalogs ASTM Standard. A919-Standard Terminology Relating to Heat Treating of Metals and ASTM Standard E-7 Standard Terminology Relating to Metallography)

*Acid Steel*—Steel melted in a furnace with an acid (siliceous) bottom and lining and under a slag which is dominantly siliceous.[G-1*]

*Age Hardening*—Hardening by aging, usually after rapid cooling or cold working.[G-2]

*Aging*—A change in the properties of certain metals and alloys that occurs at ambient or moderately elevated temperatures after hot working or a heat treatment (quench aging in ferrous alloys, natural or artificial aging in ferrous and nonferrous alloys) or after a cold-working operation (strain aging).  The change in properties is often, but not always, due to a phase change (precipitation), but never involves a change in chemical composition of the metals or alloys.[G-2]

*Alloy*—A mixture with metallic properties composed of two or more elements of which at least one is a metal.[G-1]

*Annealing*—Heating to and holding a suitable temperature and the cooling at a suitable rate, for such purposes as reducing hardness, improving machinability, facilitating cold working, producing a desired microstructure, or obtaining desired mechanical, physical, or other proprerties.[G-2]  The objects being treated are ordinarily allowed to cool slowly in the furnace. They may, however, be removed from the furnace and cooled in some medium which will prolong the time of cooling as compared to unrestricted cooling in the air.[G-1]

*Arc Burns*—Localized points of surface melting caused by arcing between an electrode or ground and pipe surface.  Arc burns are considered defects.[G-3]

*Austenite*—An important high-temperature phase of steel, the decomposition of which on cooling forms the room-temperature constituents consistent with the alloy content and the cooling rate of the sample.  It is a homogeneous phase, consisting of a solid solution of carbon in the gamma form of iron.  It is formed when steel is heated to a temperature above the transformation temperature.  The limiting temperatures for the formation of austenite vary with composition.

*Bainite*—*Upper, lower, intermediate*—Metastable microstructure or microstructures resulting from the transformation of austenite at temperatures between those which produce pearlite and martensite. If the transformation temperature is just below that at which the

---

[*]Refers to source of definition, see page G-19

finest pearlite is formed, the bainite (upper bainite) has a feathery appearance. If the transformation temperature is just above that at which martensite is produced, the bainite (lower bainite) is acicular, resembling slightly tempered martensite. Intermediate bainite resembles upper bainite; however, the carbides are smaller and more randomly oriented.[G-4]

**Banded Structure**—A structure of nearly parallel bands which run in the direction of working which results from segregation of constituents (e.g., alternating ferrite and pearlite bands in a pearlitic low carbon hot rolled steel.

**Basic Steel**—Steel melted in a furnace with a basic bottom and lining and under a slag which is dominantly basic.[G-1]

**Briggs' Standard**—A list of pipe sizes, thicknesses, threads, etc., compiled by Robert Briggs about 1862 and subsequently adopted as a standard now known as ANSI B36.10M (American National Standards Institute).[G-5]

**Bessemer Process**—A pneumatic process, developed independently by William Kelly of Eddyville, Kentucky and Henry Bessemer of England, for making steel by blowing air through a bath of molten iron contained in a bottom-blown vessel lined with acid (siliceous) refractories. The process was the first to provide a large-scale method whereby pig iron could rapidly and cheaply be refined and converted into liquid steel. Bessemer's American patent was issued in 1856; although Kelly did not apply for a patent until 1857, he was able to prove that he had worked on the idea as early as 1847. The process is one of rapid oxidation mainly of silicon and carbon.

**Blast Furnace**—A shaft furnace supplied with an air blast, usually hot, for producing pig iron by smelting iron ore. The furnace is continuous in operation, the raw materials (iron ore, coke, and limestone) are charged at the top, and the molten pig iron and slag are collected at the bottom and are tapped out at intervals.[G-5]

**Blister**—A raised spot on the surface of the pipe caused by expansion of gas in a cavity within the pipe wall.[G-3]

**Bloom**—(slab, billet, sheet bar)—Semifinished products of rectangular cross section with rounded corners, hot rolled from ingots. The chief differences are in cross sectional area, in ratio of width of thickness, and in their intended use. American Iron and Steel Institute Manual classifies general usage as follows:

| Type | Width, in. | Thickness, in. | Cross Section Area, sq.in. |
|------|-----------|----------------|----------------------------|
| Bloom | Width equals thickness | | 36 (min.) |
| Billet | 1½ (min.) | 1½ (min.) | 2¼-36 |
| Slab | 10 + (min.) | 1½ (min.) | 16 (min.) |
| Sheet Bar | 8-16 | ¼-2* | 2-32* |

*Calculated from weight range 7-54 lb. per lineal foot.

Rerolling quality blooms, slabs, and billets are intended for hot rolling into shapes, plates, strip, bars, and wire rod.

Forging quality blooms, billets, and slabs are intended for conversion into forgings.

Sheet bar is converted by rolling into sheet, black plate, and tin plate.[G-1]

**Blooming Mill**—A mill used to reduce ingots to blooms, billets, slabs, or sheet bars. Depending upon the product, the mill is called a blooming mill (cogging mill in England), a billet mill, or a slabbing mill.[G-1]

**Blowhole**—A hole produced during the solidification of metal by evolved gas which, in failing to escape, is held in pockets.[G-1]

**Burning**—The heating of a metal to temperatures sufficiently close to the melting point to cause permanent injury. Such injury may be caused by the melting of the more fusible constituents, by the penetration of gases such as oxygen into the metal with consequent reactions, or perhaps by the segregation of elements already present in the metal.[G-1]

**Butt-Welding**—A process of forming a seam by butting the white-hot edges of a bent plate together. In 1825, Cornelius Whitehouse, began manufacturing butt-welded pipe by drawing a flat plate, heated to a proper temperature, through a "bell" or die.

**Carbon Steel**—Steel which owes its properties chiefly to various percentages of carbon without substantial amounts of other alloying elements; also known as plain carbon steel.[G-1]

**Carburizing**—A process in which an austenitized ferrous material is brought into contact with a carbonaceous atmosphere of sufficient carbon potential to cause absorption of carbon at the surface and by diffusion, thus creating a concentration gradient.[G-2]

**Cementite**—A very hard and brittle compound of iron and carbon corresponding to the empirical formula $Fe_3C$. It is commonly known as iron carbide and processes an orthorhombic lattice. In "plain-carbon steels" some of the iron atoms in the cementite lattice are replace by manganese, and in "alloy steels" by other elements such as chromium or tungsten. Cementite will often appear as distinct lamellae or as spheroids or globules of varying size in hypo-eutectoid steels. Cementite is in metastable equilibrium and has a tendency to decompose into iron and graphite, although the reaction rate is very slow.[G-4]

**Chipping**—One method for removing seams and other surface defects with chisel or gouge, so that the defects will not be worked into the finished product. If the defects are removed by means of gas cutting, the term "deseaming" or "scarfing" is used. Chipping is often employed simply to remove metal apart from defects.[G-1]

**Cold Expansion**—A process used to control the final size (usually the outside diameter) of some line pipe products, particularly sizes of 18-inch and larger. The expansion may be done by internal hydraulic pressure in which case the size is limited by sets of external clamps as each length is plastically deformed by the internal pressure. Alternatively, the expansion may be done mechanically by hydraulically-operated shoes which travel along inside the pipe and expand it successively in short lengths to a predetermined limit.

**Cold Shut**—A defect produced during casting of molten metal which may result from splashing, surging, interrupted pouring, or the meeting of two streams of metal coming from different directions. It may be due to the freezing of one surface before the other metal flows over it, or to the presence of interposing surface films or dirt on cold sluggish metal, or to any factor that will prevent a prefect union where two surfaces meet that should fuse and blend.[G-1]

*Cold Weld*—A metallurgically inexact term generally indicating a lack of adequate weld bonding strength of the abutting edges, due to insufficient heat and/or pressure. A cold weld may or may not have separation in the weld line. Other more definitive terms should be used whenever possible.[G-3]

*Cold Working*—Plastic deformation of a metal at a temperature low enough to insure strain hardening. This is a process of reducing the cross-sectional area by cold rolling, cold drawing, cold extrusion, or cold forging. Cold working is employed to obtain the following effects: increased yield strength, better machinability, special size accuracy, bright surface, and the production of thinner gages than hot work can accomplish economically. Cold working usually has a detrimental effect on ductility or notch toughness and should be minimized or avoided if not done for a specific purpose.[G-1+]

*Combined Carbon*—All the carbon in iron or steel which is combined with iron or alloying elements to form carbide.[G-1]

*Contact Marks*—Intermittent marks adjacent to the weld line resulting from the electrical contact between the electrodes supplying the welding current and the pipe surface.[G-3]

*Continuous Welded Pipe*—Continuous welded pipe is pipe that has one longitudinal seam produced by the continuous welding process. (This is a type of butt-welded pipe.) [41st Edition, API 5L][G-6]

*Continuous Welding*—Continuous welding is a process of forming a seam by heating the skelp in a furnace and mechanically pressing the formed edges together wherein successive coils of skelp have been joined together to provide a continuous flow of steel for the welding mill. (This process is a type of butt welding.) [41st Edition, API 5L][G-6]

*Crack*—A stress-induced separation of the metal which, without any other influence, is insufficient in extent to cause complete rupture of the material.[G-3]

*Crop*—The end or ends of an ingot containing the pipe or other defects which are cut off and discarded; also termed "crop end" and "discard".[G-1]

*Decarburization*—The loss of carbon from the surface of a ferrous alloy as a result of heating in a medium that reacts with the carbon.[G-2]

*Defect*—An imperfection of sufficient magnitude to warrant rejection of the product. [41st Edition of API 5L][G-6]

*Dent*—A local change in surface contour caused by mechanical impact, but not accompanied by loss of metal.[G-3]

*Dissolved Carbon*—Carbon in solution in either the liquid or solid state.[G-1]

*Eccentricity*—A condition of pipe in which the outside diameter and inside diameter axes are not coincident, resulting in wall thickness variation around the circumference at a given section plane.[G-3]

*Electric Welded Pipe*—Electric welded pipe is pipe that has one longitudinal seam produced by the electric-welding process. For grades higher than X42, the weld seam and the entire heat affected zone shall be heat treated so as to simulate the normalizing heat treatment. For Grades X42 and lower, the weld seam shall be similarly heat treated or the pipe shall be processed in such a manner that no untempered martensite remains.[G-6]

*Electric Welding*—Electric welding is a process of forming a seam by electric-resistance or electric-induction welding wherein the edges to be welded are mechanically pressed together and the heat for welding is generated by the resistance to flow of the electric current. [41st Edition, API 5L][G-6]

*Elongation*—The amount of permanent extension in the vicinity of the fracture in the tension test; usually expressed as a percentage of the original gage length, such as 25 percent in 2 inches. It may also refer to the amount of extension at any stage in any process which continuously elongates a body, as in rolling.[G-1]

*Endurance Limit*—A limiting stress, below which metal will withstand without fracture an indefinitely large number of cycles of stress. If the term is used without qualification, the cycles of stress are usually such as to produce complete reversal of flexural stress. Above this limit failure occurs by the generation and growth of cracks until fracture results in the remaining section.[G-1]

**Equiaxed Grain**—A polygonal crystallite, in an aggregate, whose dimensions are approximately the same in all directions.[G-4]

**Etching**—Controlled preferential attack on a metal surface for the purpose of revealing structural details.[G-4]

**Excessive Reinforcement (Excessive Over-Fill)**—Outside weld beads which extend above the prolongation of the original surface of the pipe (more than 1/8 inch for pipe having a thickness of 1/2 inch and under, and more than 3/16 inch for a pipe having a thickness of over 1/2 inch). [API Bull. 5T1][G-3]

**Excessive Trim**—Ref. API Spec 5L — Par. 10.5

**Ferrite**—The metallographic name for nearly pure iron in steel. It exhibits a white microstructural appearance in a low-carbon steel in which it is the predominant constituent. Ferrite is also called alpha iron, and it may contain alloying elements but only in solid solution form.

**Ferroalloy**—An alloy of iron with a sufficient amount of some element or elements, such as manganese, chromium, or vanadium, for use as a means of introducing these elements into steel by admixture of molten steel. Because of their high iron content, most ferroalloys were formerly produced in the iron blast furnace. Eventually, the production of alloys for steelmaking purposes began to be carried out in electric-reduction and other types of furnaces as well, and a number of alloys now produced contain relatively little iron. For this reason, the term addition agent is preferred.[G-1,G-7]

**Finishing Temperature**—The temperature at which hot mechanical working of metal is completed.[G-1]

**Flash**—The excess metal forced to the outside and inside surface of a flash-welded or electric-welded pipe as the seam is formed by electrically induced heating and mechanical pressure. The flash is usually removed completely by trimming tools immediately following the seam welding operation. However, a distinctive outside and inside diameters portion of the flash was intentionally left untrimmed on flash-welded pipe.

**Flash-Welded Pipe**—A process of seam welding line pipe consisting of bringing together at one time both edges of a preformed can while passing an electric current between the edges and exerting a mechanical pressure to bond the locally-heated edges.  The process of flash-welding was used by only one manufacturer (A.O.Smith Corporation) to make line pipe.

*Fracture*—The irregular surface produced when a piece of metal is ruptured or broken.  The creation of new surfaces such as holes, cracks, or actual separation into two or more parts.[G-1+]

*Gas Metal-Arc Welded Pipe*—Gas metal-arc welded pipe is defined as pipe that has one longitudinal seam produced by the continuous gas metal-arc welding process.  At least one pass shall be on the inside, and at least one pass shall be on the outside.  [41st Edition, API 5L][G-6]

*Gas Metal-Arc Welding*—Gas metal-arc welding is a welding process that produces coalescence of metals by heating them with an arc or arcs between a continuous consumable electrode and the work.  Shielding is obtained entirely from an externally supplied gas or gas mixture.  Pressure is not used, and the filler metal is obtained from the electrode.  [41st Edition, API 5L][G-6]

*Gouge*—Elongated grooves or cavities caused by mechanical removal of metal.[G-3]

*Grain Boundary*—An interface separating two grains, where the orientation of the lattice changes from that of one grain to that of the other.  When the orientation change is very small, the boundary is sometimes referred to as a subboundary.[G-4]

*Grain Growth*—An increase in the grain size of metal.[G-1]

*Grains*—Individual crystals in metals.[G-1]

*Grinding (in manufacturing)*—The process of removing imperfections by mechanical (usually abrasive) means.

*Grinding (metallographic)*—The removal of material from the surface of a specimen by abrasion through the use of randomly oriented hard-abrasion particles bonded to a suitable substrate, such as paper or cloth, where the abrasive particle size is generally in the range of 60 to 600 grit (approximately 150 to 15 μm) but may be finer.[G-4]

*Hardenability*—In a ferrous alloy, the property that determines the depth and distribution of hardness induced by quenching.[G-2]

*Hardening*—Heating and quenching certain iron-base alloys from a temperature either within or above the critical temperature range for the purpose of producing a hardness superior to that obtained when the alloy is not quenched. Usually restricted to the formation of martensite.[G-1]

*Hard Spot*—An area in the pipe with a hardness level considerably higher than that of the surrounding metal, usually due to localized quenching.[G-3]

*Heat Treatment*—An operation or combination of operations involving the heating and cooling of a metal or alloy in the solid state for the purpose of obtaining certain desirable conditions or properties. Heating and cooling for the sole purpose of mechanical working are excluded from the meaning of this definition.[G-1]

*Hook Cracks or Upturned Fiber Imperfections*—Metal separations, resulting from imperfections at the edge of the plate or skelp, parallel to the surface, which turn toward the I.D. or O.D. pipe surface when the edges are upset during welding.[Bul 5TL, Electric Flash-Welds][G-3]

*Hot Shortness*—Brittleness in metal when hot.[G-1]

*Hot Working*—Plastic deformation of metal at a temperature high enough to prevent strain hardening.[G-1]

*Imperfection*—A discontinuity or irregularity in the product. [41st Edition, API 5L][G-6]

*Inadequate Flash Trim*—Detected by methods outlined in the specification. A condition in which the height of the electric weld flash after trimming exceeds the limits set in the API specification to which the pipe was manufactured.[G-3]

*Inclusion*—Foreign material or non-metallic particles entrapped within the metal during solidification.[G-3]

*Incomplete Fusion*—lack of complete coalescence of some portion of the metal in a weld joint.[G-3]

*Incomplete Penetration (Lack of Penetration)*—A condition where the weld metal does not continue through the full thickness of the joint.[G-3]

*Ingot*—A special kind of casting for subsequent rolling or forging.[G-1]

*Ingot Iron*—An open hearth iron very low in carbon, manganese, and other impurities.[G-1]

*Jointer*—A pipe trade term used to express a random length composed of two pieces welded or coupled together.[G-5]

*Killed Steel*—A steel sufficiently deoxidized to prevent gas evolution during solidification. The top surface of the ingot freezes immediately and subsequent shrinkage produces a central pipe. A semiskilled steel, having been less completely deoxidized, develops sufficient gas evolution internally in freezing to replace the pipe by a substantially equivalent volume of rather deep-seated blowholes.[G-1]

*Lamination*—An internal metal separation creating layers generally parallel to the surface.[G-3]

*Lap*—Fold of metal which has been rolled or otherwise worked against the surface of rolled metal, but has not fused into sound metal.[G-3]

*Lap Weld*—A term applied to a weld formed by lapping two pieces of metal and then pressing or hammering, particularly, to the longitudinal joint produced by a welding process for tubes or pipe in which the edges of the skelp are beveled or scarfed so that when they are overlapped they can be welded together.[G-1]

*Line Pipe*—Pipe intended for use in a pipeline.

*Longitudinal Seam Submerged-Arc Welded Pipe*—Longitudinal seam submerged-arc welded pipe is pipe that has one longitudinal seam produced by the automatic submerged-arc welding process. At least one pass is on the inside, and at least one pass is on the outside. (This type of pipe is also known as submerged-arc welded pipe.) [41st Edition, API 5L][G-6]

*Macrograph*—A graphic reproduction of an object, slightly reduced in size, unmagnified, or magnified ten diameters or less (photomacrograph).[G-4]

*Manufacturer*—A firm, company, or corporation responsible for marking the product to warrant that it conforms to the API specifications. The manufacturer may be, as applicable, a pipe mill or processor; a maker of couplings, pup joints or connectors; or a threader. The manufacturer is responsible for compliance with all of the applicable provisions of the API specifications. [41st Edition, API 5L][G-6]

*Martensite*—A microconstituent or structure in quenched steel characterized by an acicular or needle-like pattern on the surface of a polished specimen. It has the maximum hardness of any of the decomposition products of austenite.[G-1]

*Mechanical Properties*—Those properties that reveal the reaction, elastic and inelastic, of a material to an applied force, or that involve the relationship between stress and strain (e.g., Young's modulus, tensile strength, fatigue limit).[G-1]

*Mechanical Testing*—Testing methods by which mechanical properties are determined.[G-1]

*Mechanical Working*—Subjecting metal to pressure exerted by rolls, presses, or hammers, to change its form, or to affect the structure and therefore the mechanical and physical properties.[G-1]

*Metallography*—That branch of science which relates to the constitution and structure, and their relation to the properties, of metals and alloys.[G-4]

*Micrograph*—A graphic reproduction of an object as seen through the microscope or equivalent optical instrument, at a magnification greater than ten diameters (photomicrograph).[G-4]

*Microstructure*—The structure of a suitably prepared specimen as revealed by a microscope.[G-4]

*Modulus of Elasticity*—The ratio, within the limit of elasticity, of the stress to the corresponding strain. The stress in pounds per square inch is divided by the elongation in fractions of an inch for each inch of the original gage length of the specimen.[G-1]

*Normalizing*—Heating a ferrous alloy to a suitable temperature above the transformation range and then cooling in air to a temperature substantially below the transformation range.[G-2]

*Offset of Plate Edges*—The radial offset of plate edges in weld seams.[G-3]

*Out-of-Line Weld Beads or OFF Seam*—A condition in deposited welds when the inner and/or outer weld beads are sufficiently out of radial alignment with the abutting edges of the joint to cause incomplete penetration.[G-3]

*Pearlite*—The lamellar aggregate of ferrite and carbide resulting from the direct transformation of austenite at temperatures below the $Ar_3$.

*Penetrator*—A localized spot of incomplete fusion in electric welded pipe.[G-3]

*Piercing*—Producing a hole in metal by forcing an instrument through it. Usually refers to making steel tubes from solid steel bars.[G-1]

*Pig Iron*—Pig Iron is the product of the blast furnace and is made by the reduction of iron ore. With the exception of direct metal, pig iron is either remelted and cast (producing gray and white iron), or it is refined in the steel making processes.[G-1]

*Pinhole*—A short unwelded area in the weld line extending through the entire pipe thickness so that fluid will leak out through the area very slowly.[G-3]

*Pipe*—A cavity formed in metal (specially ingots) during the solidification of the last portion of liquid metal. Contraction of the metal causes this cavity or pipe.[G-1]

*Pipe Mill*—A firm, company, or corporation that operates pipe making facilities.[G-6]

*Pit*—A sharp depression in the surface of metal resulting from the removal of foreign material rolled into the surface during manufacture.

**Plain End**—Usually contracted to P.E.—used to signify pipe cut off and not threaded, i.e., ends left as cut square or beveled.[G-5]

**Plug Scores**—Internal longitudinal grooves occurring in seamless pipe, usually caused by hard pieces of metal adhering to the high-mill plug.[G-3]

**Polishing**—A mechanical, chemical, or electrolytic process or combination thereof used to prepare a smooth reflective surface suitable for microstructure examination, free of artifacts or damage introduced during prior sectioning or grinding.[G-4]

**Porosity**—Voids in a metal, usually resulting from shrinkage or gas entrapment occurring during solidification of a casting or deposited weld.[G-3]

**Precipitation Hardening**—Hardening caused by the precipitation of a constituent from a supersaturated solid solution.[G-2]

**Processor**—A firm, company, or corporation that operates facilities capable of heat treating pipe made by a pipe mill.[G-6]

**Quench Hardening**—Hardening a ferrous alloy by austenitizing and then cooling rapidly enough so that some or all of the austenite transforms to martensite.[G-2]

**Quenching**—Rapid cooling by immersion in liquid or gases or by contact with metal.[G-1]

**Random Length**—The "catch length" or length of good quality pipe after its ends have been trimmed. For Butt and Lap Weld pipes, this was usually about 20 feet or less. Thus, 40-foot pieces or "joints" are referred to as double random lengths, 60-foot joints are triple random lengths, etc.[G-5]

**Reduction in Area**—The difference between the original cross sectional area and that of the smallest area at the point of rupture. It is usually stated as a percentage of the original area.[G-1]

**Rimmed Steel**—An incompletely deoxidized steel normally containing less than 0.25 percent carbon and having the following characteristics: (a) During solidification an evolution of gas occurs sufficient to maintain a liquid ingot top ("open" steel) until a side and bottom rim of substantial thickness has formed. if the rimming action is intentionally stopped shortly after the mold is filled the product is termed *capped steel*. (b) After complete solidification, the ingot consists of two distinct zones: A rim somewhat purer than when poured and a core containing scattered blowholes with a minimum amount of pipe and having an average inclusion content somewhat higher than when poured and markedly higher in the upper portion of the ingot.[G-1]

**Rolled-in Slugs**—A foreign metallic body rolled into the metal surface, usually not fused.[G-3]

**Roll Mark**—A term applied to surface imperfections caused by improper roll alignment or roll surface damage. Such imperfections may be periodic or continuous.[G-3]

**Scab**—An imperfection in the form of a shell or veneer, generally attached to the surface by sound metal. It usually has its origin in an ingot defect.[G-3]

**Seam**—Crevice in rolled metal which has been more or less closed by rolling or other work but has not been fused into sound metal.[G-3]

**Seamless Process**—The process is a method of hot working steel to form a tubular product without a welded seam. If necessary, the hot worked tubular product may be subsequently cold finished to produce the desired shape, dimensions, and properties. [41st Edition, API 5L][G-6]

**Segregation**—Concentration of alloying elements in specific regions in a metallic object.[G-4]

**Skelp**—The flat rolled steel form from which welded pipe is produced.

**Slag Inclusions**—Non-metallic solid material entrapped in the weld deposit or between weld metal and base metal.[G-3]

**Sliver**—An extremely thin elongated piece of metal that has been rolled into the surface of the parent metal to which it is attached usually at only one end.[G-3]

**Soaking**—Holding steel at an elevated temperature for the attainment of uniform temperature throughout the piece.[G-1]

*Spellerizing*—The method of treating metal, which consists in subjecting the heated bloom to the action of rolls having regularly shaped projections on their working surfaces, then subjecting the bloom to the action of smooth faced rolls, and repeating the operation, whereby the surface of the metal is worked to produce a uniformly dense texture, believed to be better adapted to resist corrosion, especially in the form of pitting.[G-5]

*Spheroidizing*—Heating and cooling to produce a spheroidal or globular form of carbide in steel. Spheroidizing methods frequently used are:
1. Prolonged holding at a temperature just below $Ae_1$.
2. Heating and cooling alternately between temperatures that are just above and just below $Ae_1$.
3. Heating to a temperature above $Ae_1$ or $Ae_3$ and then cooling very slowly in the furnace or holding at a temperature just below $Ae_1$.
4. Cooling at a suitable rate from the minimum temperature at which all carbide is dissolved, to prevent the re-formation of a carbide network, and then reheating in accordance with method (1) or (2) above. (Applicable to hypereutectoid steel containing a carbide network).[G-2]

*Standard Pipe*—Standard pipe is welded or seamless pipe commonly produced in three classes of wall thickness: standard weight, extra strong and double extra strong.[G-8]

*Stitching*—Variation in the properties of an electric weld occurring at short regular intervals along the weld line due to repetitive variation in welding heat. The variation in properties gives rise to a regular pattern of light and dark areas visible only when the weld is broken in the weld line.[G-3]

*Stress Relieving*—Heating to a suitable temperature, holding long enough to reduce residual stresses and then cooling slowly enough to minimize the development of new residual stresses.[G-2]

*Stretch Mill Indentation*—Localized thinning of the pipe body wall—usually located on the inside surface of pipe which has been stretch reduced.[G-3]

*Stringer*—A single, high-aspect ratio, elongated inclusion, two or more elongated inclusions, or a number of small non-deformable inclusions aligned in a linear pattern due to deformation.[G-4]

*Submerged-Arc Welding*—Submerged-arc welding is a welding process that produces coalescence of metals by heating them with an arc or arcs between a bare metal consumable electrode or electrodes and the work. The arc and molten metal are shielded by a blanket of granular, fusible material on the work. Pressure is not used, and part or all of the filler metal is obtained from the electrode or electrodes. [41st Edition, API 5L][G-6]

*Surface Hardening*—A generic term covering several processes applicable to a suitable ferrous alloy that produces by quench hardening only, a surface layer that is harder or more wear resistant than the core. There is no significant alternation of the chemical composition of the surface layer. The processes commonly used are induction hardening, flame hardening, and shell hardening. Use of the applicable specific process name is preferred.[G-2]

*Teeming*—See pouring. Usually refers to pouring of metal into molds.[G-1]

*Tempered Martensite*—The decomposition products which result from heating martensite to temperatures below the ferrite austenite ($Ae_1$) transformation temperature.[G-4]

*Tempering*—(1) Reheating a quench hardened or normalized ferrous alloy to a temperature below the transformation range ($Ac_1$), and then cooling at any desired rate. (2) A term used in conjunction with a qualifying adjective to designate the relative properties of a particular metal or alloy induced by cold work or heat treatment, or both.[G-2]

*Transformation Ranges* or *Transformation Temperature Ranges*—Those ranges of temperature within which austenite forms during heating and transforms during cooling. The two ranges are distinct, sometimes overlapping but never coinciding. The limiting temperatures of the ranges depend on the composition of the alloy and on the rate of change of temperature, particularly during cooling.[G-2]

*Transformation Temperature*—The temperature at which a change in phase occurs. The term is sometimes used to denote the limiting temperature of a transformation range. The following symbols are used for iron and steels:

$Ac_{cm}$—In hypereutectoid steel, the temperature at which the solution of cementite in austenite is completed during heating.

$Ac_1$—The temperature at which austenite begins to form during heating.

$Ac_3$—The temperature at which transformation of ferrite to austenite is completed during heating.

$Ac_4$—The temperature at which austenite transforms to delta ferrite during heating.

$Ae_1$, $Ae_3$, $Ae_{cm}$, $Ae_4$—The temperatures of phase changes at equilibrium.

$Ar_{cm}$—In hypereutectoid steel, the temperature at which precipitation of cementite starts during cooling.

$Ar_1$—The temperature at which transformation of austenite to ferrite or to ferrite plus cementite is completed during cooling.

$Ar_3$—The temperature at which austenite begins to transform to ferrite during cooling.

$Ar_4$—The temperature at which delta ferrite transforms to austenite during cooling.

$M_f$—The temperature, during cooling, at which transformation of austenite to martensite is substantially completed.

$M_s$—The temperature at which transformation of austenite to martensite starts during cooling.

NOTE—All these changes except the formation of martensite occur at lower temperatures during cooling than during heating, and depend on the rate of change of temperature.

**Under-Cut**—Under-cutting on submerged-arc-welded pipe is the reduction in thickness of the pipe wall adjacent to the weld where it is fused to the surface of the pipe.[G-3]

**Upset Underfill**—A depression on the outside or inside surface of an upset caused by insufficient flow of metal to completely fill out the upset to the desired shape.[G-3]

**Upset Wrinkles**—Surface irregularity occurring on pipe upsets in the form of transverse forging laps.[G-3]

**Weld Area Crack**—A crack that occurs in the weld deposit, the fusion line, or the heat affected zone. (**Crack:** A stress-induced separation of the metal which, without any other influence, is insufficient in extent to cause complete rupture of the material.)[G-3]

**Woody Fracture**—A descriptive term for fracture of sound though dirty steel, frequently also reedy or conchoidal in appearance, and often containing discernible slag particles. Woody fractures sometimes contain many small silvery areas, too numerous and small to be correctly termed "flakes", and of a different nature.[G-3]

**Work Hardening**—A change in the hardness of a material as a result of plastic deformation.[G-4]

**Yield Strength**—Yield strength is the stress at which a material exhibits a specified limited deviation from proportionality of stress to strain. The deviation is expressed in terms of strain, percent offset, total extension under load, etc. (ASTM A370-95)

## References

G-1. Pittsburgh Steel's 1944 Catalog, "The Making, Shaping and Treating of Steel"

G-2. A919-Standard Terminology Relating to Heat Treating of Metals

G-3. API Bulletin 5T1, Ninth Edition, May 31, 1988

G-4. ASTM Standard E-7 Standard Terminology Relating to Metallography)

G-5. National Tube Company Catalogs ASTM Standard

G-6. API Specification 5L, Various Editions

G-7. ASM Metals Handbook, Various Editions

G-8. AISI

www.ingramcontent.com/pod-product-compliance
Lightning Source LLC
Chambersburg PA
CBHW061346210326
41598CB00035B/5894